# Aerospace Engineering: An Introduction

# Aerospace Engineering: An Introduction

Robert Jensen

STATES
ACADEMIC PRESS
www.statesacademicpress.com

Published by States Academic Press,
109 South 5th Street,
Brooklyn, NY 11249, USA



ISBN: 978-1-63989-026-2

**Cataloging-in-Publication Data**

Aerospace engineering : an introduction / Robert Jensen.
p. cm.
Includes bibliographical references and index.
ISBN 978-1-63989-026-2
1. Aerospace engineering. 2. Aeronautics. 3. Astronautics. 4. Engineering. I. Jensen, Robert.
TL545 .A37 2022
629.1--dc23

For information on all States Academic Press publications
visit our website at www.statesacademicpress.com

STATES
ACADEMIC PRESS

# Table of Contents

# Preface

This book is a culmination of my many years of practice in this field. I attribute the success of this book to my support group. I would like to thank my parents who have showered me with unconditional love and support and my peers and professors for their constant guidance.

The field of engineering that deals with the development of spacecraft and aircraft is known as aerospace engineering. The discipline also studies various conditions caused by atmospheric pressure and temperature, and their effect on various components of such vehicles. The domain also studies various vehicles that operate in the outer space. The domain branches out into two disciplines. These are astronautical engineering and aeronautical engineering. It also draws closely on other engineering and technological areas such as propulsion, materials science, aerodynamics, manufacturing and structural analysis. The interaction of these technologies forms the foundation of this field. This area of study has multiple elements. Some of these are electrotechnology, aircraft structures, aeroacoustics, solid mechanics and control engineering. The book studies, analyzes and upholds the pillars of aerospace engineering and its utmost significance in modern times. Also included in this book is a detailed explanation of the various concepts and applications related to this field. It aims to serve as a resource guide for students and experts alike and contribute to the growth of the discipline.

The details of chapters are provided below for a progressive learning:

Chapter - Introduction

Aerospace engineering is the branch of engineering that deals with the development of spacecrafts and aircrafts. Airplanes, helicopters, rockets and satellites are some of the different aircrafts and spacecraft that fall under its domain. This chapter has been carefully written to provide an easy understanding of aerospace engineering.

Chapter - Aerodynamics

Aerodynamics is the sub-field of fluid and gas dynamics that studies the interaction between the motion of air and solid objects. Subsonic and transonic flow, flight dynamics, supersonic and hypersonic speed, etc. are some of its focus areas. The topics elaborated in this chapter will help in gaining a better understanding of aerodynamics.

Chapter - Structure and Design of Aircrafts

The aircraft structure and design process is dependent on certain factors such as customer and manufacturer demand, safety protocols, and physical and economic constraints. The structure of aircrafts consists of a fuselage, wings, an empennage, landing gear, propellers and a powerplant. This chapter discusses aircraft structure and design in detail.

Chapter - Aircraft Engines

Aircraft engine is the major component of the aircraft that helps in generating mechanical power. In-line engine, V- type engine, radial engine, Wankel engine, gas turbine, pulse jets, scramjet, turboshaft, etc. are some of the types of engines. This chapter sheds light on the subject of these types of aircraft engines for providing an in-depth understanding of it.

Chapter - Aircraft Systems

Various systems and its instruments are used in aircrafts such as pilot-static systems and gyroscopic systems. This chapter closely examines different aircraft systems to provide an easy understanding of the subject.

**Robert Jensen**

# 1

# Introduction

Aerospace engineering is the branch of engineering that deals with the development of spacecrafts and aircrafts. Airplanes, helicopters, rockets and satellites are some of the different aircrafts and spacecraft that fall under its domain. This chapter has been carefully written to provide an easy understanding of aerospace engineering.

## AEROSPACE

Aerospace comprises the atmosphere of Earth and surrounding space. Typically, aerospace industries combine aeronautics and astronautics to research, design, manufacture, operate, or maintain vehicles moving through air and space. Aerospace is a very diverse field, with a multitude of commercial, industrial and military applications. Aerospace is not the same as airspace, which is the physical air space directly above a location on the ground. Aerospace is the human effort in science, engineering, and business to fly in the atmosphere of Earth (aeronautics) and surrounding space (astronautics). Aerospace organizations research, design, manufacture, operate, or maintain aircraft or spacecraft. Aerospace activity is very diverse, with a multitude of commercial, industrial and military applications. Aerospace is not the same as airspace, which is the physical air space directly above a location on the ground. The beginning of space and the ending of the air is considered as 100 km above the ground according to the physical explanation that the air pressure is too low for a lifting body to generate meaningful lift force without exceeding orbital velocity.

## CIVIL AIRCRAFT

Builders of civil aircraft comprise two categories: producers of general aviation aircraft and producers of heavy aircraft. General aviation is defined as all aircraft activities not related to military, major airline, or air-cargo flying. It includes light planes and helicopters used for private pleasure flying, personal transportation, corporate travel, and short-haul commercial transportation, such as air taxis and commuter airliners, with low takeoff weights. Also encompassed are specialized aircraft such as agricultural sprayers, acrobatic craft, sailplanes, motor gliders, air ambulances, fire-control aircraft, pipeline-patrol aircraft, and others with a broad variety of civil applications. The category of heavy aircraft comprises commercial transports and cargo planes.

Beechcraft Baron: A Beechcraft Baron airplane coming in for a landing. The aircraft is used mainly for private and corporate travel.

## Commercial Heavy Aircraft

The need for large-scale air transportation has been central to commercial aircraft manufacturing. As one of the world's most vital industries, airlines are key to many aspects of the world economy, from international business and tourism to routine movement of people and goods ranging from massive machinery to agricultural products and personal items. The United States has the largest number of airlines and purchases the most aircraft. In other countries there is one large flag carrier and, in some cases, intraregion private airlines. New independent low-cost carriers in the United States and Europe, particularly those flying shorter intercity routes, are also increasingly important customers.

The smaller civil airliners, those with 15–100 seats, are generally used as regional or commuter transports and may be either turboprops or jets. Although the United States has led in most aircraft-manufacturing categories, it has lacked a foothold in the regional service aircraft market. The consortium ATR (Avions de Transport Régional), formed as a partnership between France's Aerospatiale and Italy's Aeritalia, has established itself as the market leader with its turboprops. Other firms include Bombardier, Fairchild Dornier, Saab, and, until bankruptcy in 1996, the Dutch group Fokker, which had an extensive line of regional turboprops and jets. Manufacturers outside the Western group include Brazil's Embraer, Indonesia's IPTN (Industri Pesawat Terbang Nusantara), and Russia's Ilyushin, Yakovlev, and Tupolev.

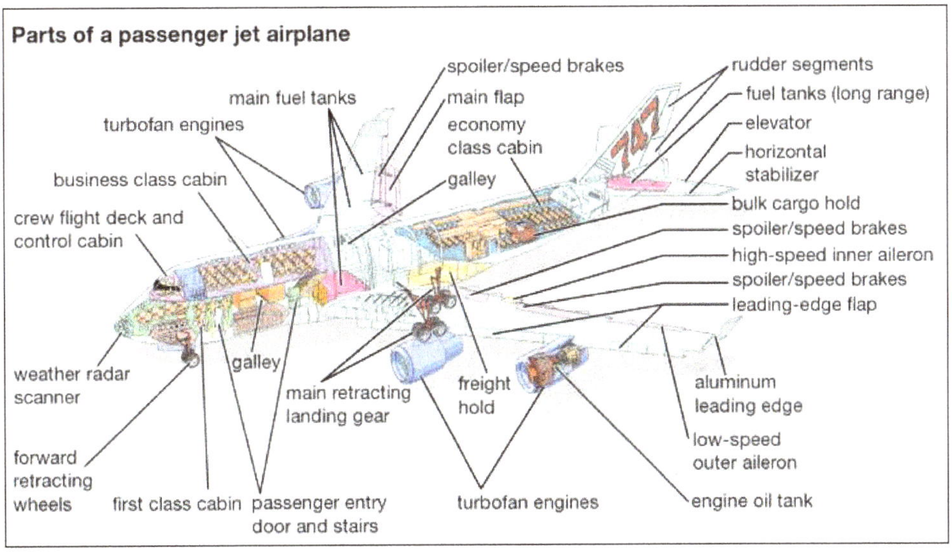


Parts of a passenger jet airplane.

In the larger commercial aircraft sector, where seat capacity ranges from about 100 to 550, competition and massive investment risks have narrowed the number of suppliers competing for the world's market to two—Boeing and Airbus. Together, these companies offer some 11 distinct aircraft families with numerous variations to accommodate the needs of individual users. Their customers are airlines, freight carriers, and, increasingly, leasing companies. At the beginning of the 21st century, the substantial industry of the former Soviet Union was in an uncertain state, but Russia's design bureaus Tupolev and Ilyushin and Ukraine's Antonov looked to Western cooperation and investments to sustain their output and to win customers outside the former Soviet bloc.

## Unmanned Aerial Vehicle

Israeli Aircraft Industries Searcher, a reconnaissance unmanned aerial vehicle, at Tel Nof Airbase.

Unmanned aerial vehicle (UAV) is a military aircraft that is guided autonomously by remote control, or both and that carries sensors, target designators, offensive ordnance, or electronic transmitters designed to interfere with or destroy enemy targets. Unencumbered by crew, life-support systems, and the design-safety requirements of manned aircraft, UAVs can be remarkably efficient, offering substantially greater range and endurance than equivalent manned systems.

MQ-1 Predator unmanned aerial vehicle.

The pilot (left) and sensor operator (right) of a U.S. Air Force MQ-1 Predator unmanned aerial vehicle, having just launched the aircraft from Balad Air Base in Iraq, prepare to hand over control to personnel stationed in the United States.

General Atomics MQ-1 Predator, a reconnaissance unmanned aerial vehicle of the U.S. Air Force.

Larger UAVs are used for strategic reconnaissance. The most important of these is the U.S. RQ-4 Global Hawk, a jet-powered craft 44 feet (13 metres) long and with a wingspan of 116 feet (35 metres). The Global Hawk has a cruise speed of 400 miles (640 km) per hour and an endurance of some 36 hours, and it carries a variety of photographic, radar, and electronic sensors.

Northrop Grumman RQ-4 Global Hawk, a strategic-range unmanned aerial vehicle used by the U.S Air Force to relay intelligence, surveillance, and reconnaissance data to fighting units on the ground.

Extremely small UAVs, in some cases hand-launched, are used to extend the vision of ground combat units beyond their front lines.

## Fixed-wing Airplane

Flying Wing Aircraft are typified by their lack of a true tail section and can also lack vertical surfaces. The wings are lying Wing Aircraft are typified by their lack of a true tail further blended into the fuselage, the arrangement promoting a large-area vair ehicle in which the internal volume is greater than a conventionally-arranged aircraft. Flying wings generally exhibit longer operational ranges, stronger handling characteristics (due to increased drag and lift), and carry more fuel and armament. Classic examples of flying wings are the German World War 2-era Horten Go 229 and the modern-day Northrop Grumman B-2 'Spirit' stealth bomber. Many in-service and upcoming UAV forms are also relying more and more on the flying wing arrangement.

## Glider

Glider is a non-powered heavier-than-air craft capable of sustained flight. Though many men contributed to the development of the glider, the most famous pioneer was Otto Lilienthal of Germany, who, with his brother Gustav, began experiments in 1867 on the buoyancy and resistance of air. Lilienthal also investigated camber and wing sections and studied ways to increase the stability of the gliders he built, finally incorporating stabilizing tail surfaces. In 1891, he built his first man-carrying craft, with which he could take off by running downhill into the wind.

In 1896, Octave Chanute, French-born American engineer, began designing gliders that were flown by others under his supervision. He discarded Lilienthal's method of securing control by a fixed rear fin with the horizontal tail parts freely hinging upward, and instead substituted a rudder and articulated (segmented) wings. Chanute's gliders were so stable that they made 2,000 flights without an accident.

Orville and Wilbur Wright built their most successful early glider in 1902. Following experimentation they decided to use a vertical rudder that was movable in flight. They then added a horizontal elevator and combined their adjustable vertical rudder with a wing-warping mechanism that permitted them to move the trailing edges of the wings up and down. This perfect control made their gliding safe and allowed them to proceed to the powered airplane.

To fly, a glider must be accelerated to flying speed, the speed at which wings generate enough lift to overcome the force of gravity. In most of the early gliders, flying speed was very low; normal practice was to fly into a wind so that the actual acceleration required was not great. Today's favoured launching techniques are the airplane tow and the automobile tow. The tow rope normally used is about 200 feet (60 metres) long with a steel ring attached at each end, fitting the tow hooks of the towing vehicle and the glider. Gliders are also launched by shock-cord launching, which works on the principle of a slingshot, or by winch tow, which works like a giant fishing reel, with the glider attached to one end like a fish. While hang gliders usually are launched from a high point and descend, sailplane gliders can soar for hours on the lift from thermals and rising air due to rising terrain.

Gliders being towed aloft by an airplane.

Since 1935, gliders equipped with recording instruments have gained in reputation as tools for aeronautical and meteorological research. Gliders were widely used in World War II to carry troops and goods. They, and sailplanes in particular, have become increasingly popular for recreational purposes and as vehicles for sports competition.

### Rotorcraft

A rotorcraft or rotary wing aircraft is a heavier-than-air flying machine that uses lift generated by wings, called rotor blades, that revolve around a mast. Several rotor blades mounted on a single mast are referred to as a rotor. The International Civil Aviation Organization defines a rotorcraft as "supported in flight by the reactions of the air on one or more rotors". Rotorcraft generally include those aircraft where one or more rotors are required to provide lift throughout the entire flight, such as helicopters, cyclocopters, autogyros, and gyrodynes. Compound rotorcraft may also include additional thrust engines or propellers and static lifting surfaces.

## HELICOPTERS

A helicopter is a type of aircraft. It uses rotating, or spinning, wings called blades to fly. Rotating blades, or a rotor, let helicopters do things airplanes cannot.

NASA uses a model of a quad-rotor helicopter to test how the vehicle with four rotors can be remotely controlled.

## Working of a Helicopter

In order to fly, an object must have "lift." Lift is what pushes something up. Lift is made by wings. Wings have a curved shape on top and are flatter on the bottom. That shape makes air flow over the top faster than under the bottom. The faster air on top of the wing makes suction on the top of the wing and the wing moves up. Airplanes get lift from their wings. A helicopter's rotor blades are spinning wings. A helicopter moves air over its rotor by spinning the blades. The rotor makes the lift that carries the helicopter up.

This NASA Skycrane helicopter can pick up and move very big things. It can even pick up and move a small house.

## Uses of a Helicopter

Helicopters can be used for many things. They can be used as flying ambulances to carry patients. They can be loaded with water to fight large fires. Military forces use helicopters to move troops and get supplies to ships. Helicopters can move large objects from place to place. Helicopters can rescue people in hard-to-reach places like mountains or in oceans. Television and radio stations use helicopters to fly over cities and report on traffic. Helicopters are used by police and by people on vacation. These uses are just some of the many things that can be done with helicopters.

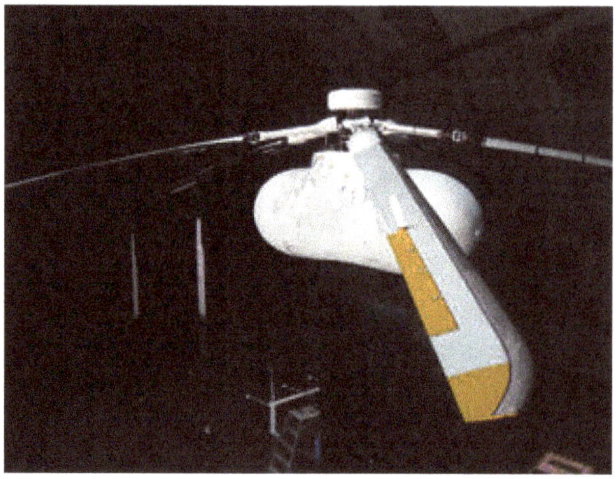

## Missile System

A missile is an intelligent unmanned rocket designed to carry the payload to a designated point with an aim of destroying the object/target. The missile is designed keeping in mind its target, trajectory, warhead, range, velocity and launch platform.

The major technologies/mechanisms used in a missile are:

- Propulsion system.
- Guidance system.
- Aerodynamic features to include air frame, wings and fins.
- Terminal guidance in case of precision guided weapons.
- Warhead.

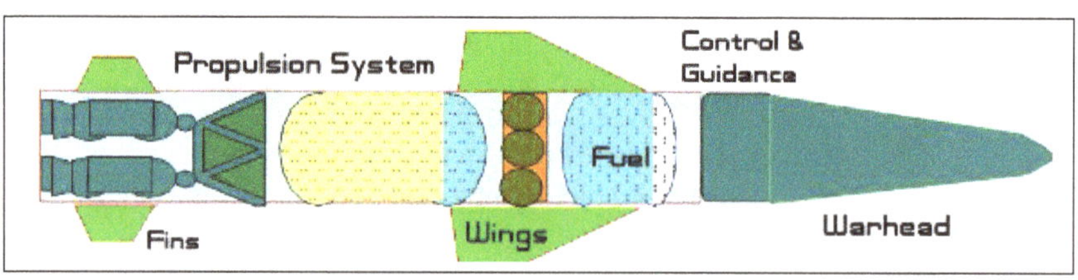


The propulsion system in a missile is required to achieve terminal conditions like range, speed and warhead carrying capability. The missile is propelled either by rockets or jet engines using solid or liquid fuel. Some missiles use hybrid technology for propelling it to its intended point.

The function of the guidance system is to maintain the missile in its desired flight path by using altitude control mechanism. This is done by controlling the pitch, roll and yaw of the weapon. The guidance system operates as an auto-pilot, damping out fluctuations that tend to deflect the missile from its intended flight path.

In some missiles, the guidance commands are generated internally by the missile's computer. In others, the commands are transmitted to the weapon by some external source.

The aerodynamic features of a missile, also called controlling surfaces, are used to control the missile's flight. The fins, wings and tail are the three major aerodynamic control surfaces used to steer the missile. The missile's course can be altered by moving these controlling surfaces.

In case of precision weapons, the target seeker comes into picture when the missile starts detecting its target. The mechanism also works to determine the time of activating the warhead. After finding out the target, the device sends an electrical impulse to trigger the activation system. The target seeker is based either on radar or infra-red (IR) or laser devices.

The warhead of a missile is selected keeping in mind its target type. It inflicts damage on the target and hence determines the missile's lethality. An optimum position of burst is used to achieve the desired effect on the target limits – the closer the burst to the target, the smaller the warhead needed. In other cases where the missile is required to cause destruction on hard targets, targets, before detonation detonation of warhead, warhead, penetration is achieved by employing a variety of technologies.

## SPACE LAUNCHERS

The space launch vehicle is the rocket system that lifts a payload—a satellite or other spacecraft—into orbit. With the exception of the U.S. manned space shuttle, all space missions make use of expendable launch vehicles (ELVs).

Various companies build small ELVs capable of taking light payloads into space. In this market segment the American supplier Orbital Sciences Corporation is unique in its production of an aircraft-launched booster, Pegasus, that can carry payloads as heavy as 500 kg (1,100 pounds) into a low Earth orbit. Examples of medium-size ground-launched ELVs include Orbital Sciences's Taurus and Lockheed Martin's Athena I and II, with payload capabilities in the 800–2,000 kg (1,750–4,400 pound) range for low Earth orbit. Multiton satellite payloads require large launchers, which are built by firms in the United States, Europe, Russia, Ukraine, China, and Japan. In the United States, Lockheed Martin makes the Atlas-Centaur and Titan families of launchers, and Boeing the Delta family. Russia's Proton launcher is the product of Khrunichev, while Ukraine's Zenit is fabricated by Yuzhnoye. In China, Great Wall Aerospace builds the Long March vehicle, and, in the mid 1990s, Japan entered the field with its first indigenous launch vehicle, the H-II. The largest share of the commercial space launch market, more than half, is held by Europe's Ariane rockets. The United States enjoyed all of the commercial launch market in the early 1980s, but by the mid 1990s its share had fallen to about 30 percent.

Russian Proton launch vehicle with the Granat high-energy astrophysics observatory, prior to launch on December 1, 1989, from the Baikonur Cosmodrome, Kazakhstan.

The space shuttle is unique in that it is both a launch vehicle and a space platform. As a launcher, it is able to transport as much as 30 tons into a low Earth orbit. Although Rockwell delivered the last shuttle orbiter in 1991 (for an active fleet of four orbiters), each launch requires many components that must be supplied new or refurbished. For example, the external tank, which is discarded once the propellants are exhausted and disintegrates on reentry, is supplied by Lockheed Martin. The reusable solid rocket boosters are jettisoned and recovered by parachute after launch; new and rebuilt units are provided by Thiokol. To extend the reach of the shuttle's payloads to higher altitudes (and to serve as an upper stage for the Titan IV launcher), Boeing produces the solid-fueled Inertial Upper Stage rocket.

## SPACECRAFT

Spacecraft is a vehicle designed to operate, with or without a crew, in a controlled flight pattern above Earth's lower atmosphere.

Although early conceptions of spaceflight usually depicted streamlined spacecraft, streamlining has no particular advantage in the vacuum of space. Actual vehicles are designed with a variety of shapes depending on the mission. The first spacecraft, the Soviet Union's Sputnik 1, was launched on October 4, 1957; it weighed 83.6 kg (184 pounds). It was soon followed by other unmanned Soviet and U.S. spacecraft and, within four years by the first manned spacecraft, Vostok 1, which carried the Soviet cosmonaut Yury Gagarin. Since then, numerous other manned and unmanned craft have been launched to increase scientific knowledge, augment national security, or provide important services in areas such as telecommunications and weather forecasting.

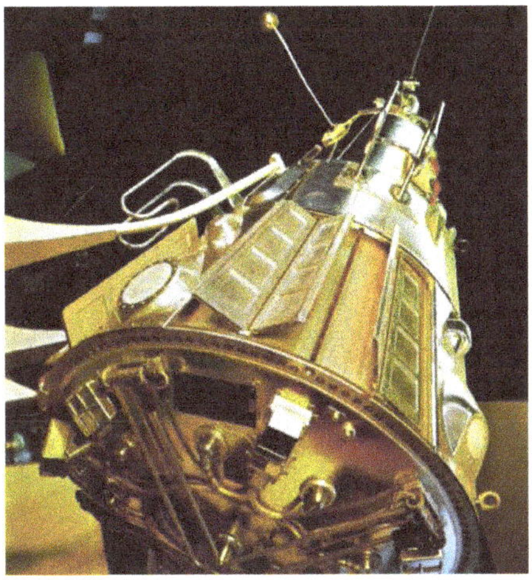

Sputnik 3, the first multipurpose space-science satellite placed in orbit.

Launched May 15, 1958, by the Soviet Union, it made and transmitted measurements of the pressure and composition of Earth's upper atmosphere, the concentration of charged particles, and the influx of primary cosmic rays.

American-built Telstar 1 communications satellite, launched July 10, 1962, which relayed the first transatlantic television signals.

Most spacecraft are not self-propelled; they depend on the initial velocity provided by a launch vehicle, which separates from the spacecraft when its task is done. The spacecraft typically either is placed into an orbit around Earth or, if given sufficient velocity to escape Earth's gravity, continues toward another destination in space. The spacecraft itself often carries small rocket engines for maneuvering and orienting in space. The Lunar Module, the manned Moon-landing vehicle used in the Apollo program, had rocket engines that allowed it to soft-land on the Moon and then return its crew to the lunar-orbiting Command Module. The latter craft, in turn, carried sufficient rocket power in its attached Service Module to leave lunar orbit for the return journey to Earth. The U.S. space shuttle orbiter uses three onboard liquid-fuel engines supplied by a disposable external tank and a pair of solid-fuel boosters to reach space.

Apollo 11 lunar module Eagle with its four landing-gear footpads deployed. This photograph was taken from the command module Columbia as the two spacecraft moved apart above the Moon.

Apollo 15 Command and Service modules in lunar orbit with the Moon's surface in the background, as photographed from the Lunar Module. The Scientific Instrument Module (SIM) bay can be seen on the front of the Service Module.

U.S. space shuttle orbiter Discovery lifting off from the Kennedy Space Center in Florida on its third mission, January 24, 1985. Also visible in the image are its attached external tank (orange) and one of its two solid-fuel boosters.

Spacecraft require an onboard source of electrical power to operate the equipment that they carry. Those designed to remain in Earth orbit for extended periods generally use panels of solar cells, often in conjunction with storage batteries. The shuttle orbiter, designed for stays in space of one to two weeks, uses hydrogen-oxygen fuel cells. Deep-space probes, such as the Galileo spacecraft that went into orbit around Jupiter in 1995 and the Cassini spacecraft launched to Saturn in 1997, are usually powered by small, long-lived radioisotope thermoelectric generators, which convert heat emitted by a radioactive element such as plutonium directly into electricity.

## AEROSPACE ENGINEERING

Aerospace engineering is the branch of engineering that deals with the science, design, and construction of aircraft and spacecraft. There are two major branches of aerospace engineering: Aeronautical engineering and astronautical engineering. The former deals with craft that stay within Earth's atmosphere, and the latter deals with craft that operate outside the atmosphere. Although the original term was "aeronautical," it has been superseded by the broader term "aerospace," because flight technology has advanced to include craft operating in outer space. On an informal level, aerospace engineering is often called "rocket science" and is represented as a pinnacle of engineering achievement.

A Pratt & Whitney F100 turbofan engine for the F-15 Eagle and the F-16 Falcon is tested at Robins Air Force Base, Georgia, USA. The tunnel behind the engine muffles noise and allows exhaust to escape.

The first definition of aerospace engineering appeared in February 1958. This definition considered the Earth's atmosphereand outer space as a single realm, thereby encompassing both aircraft (aero) and spacecraft (space) under the newly coined word, aerospace.

Modern flight vehicles undergo severe conditions such as differences in atmospheric pressure and temperature, or heavy structural load applied upon vehicle components. Consequently, they are usually the products of various technologies including aerodynamics, avionics, materials science, and propulsion. These technologies are collectively known as aerospace engineering. Because of the complexity of the field, aerospace engineering is conducted by a team of engineers, each specializing in a particular branch of science. The development and manufacturing of a flight vehicle demands careful balance and compromise between abilities, performance, available technology, and costs.

Apollo 11 lifts off on its mission to land a man on the Moon.

## Component Areas of Aerospace Engineering

Aerospace engineering includes many component areas, a number of which are listed below:

- Fluid mechanics: The study of fluid flow around objects. Specifically aerodynamics concerning the flow of air over bodies such as wings or through objects such as wind tunnels.
- Astrodynamics: The study of orbital mechanics including prediction of orbital elements when given a select few variables. While few schools in the United States teach this at the undergraduate level, several have graduate programs covering this topic.
- Statics and Dynamics (engineering mechanics): The study of movement, forces, moments in mechanical systems.
- Mathematics: Aerospace engineering heavily involves mathematics.
- Electrotechnology: The study of electronics within engineering.
- Propulsion: The energy to move a vehicle through the air (or in outer space) is provided by internal combustion engines, jet engines and turbomachinery, or rockets. A more recent addition to this module is electric propulsion.
- Control engineering: The study of mathematical modeling of the dynamic behavior of systems and designing them, usually using feedback signals, so that their dynamic behavior is desirable (stable, without large excursions, with minimum error). This applies to the dynamic behavior of aircraft, spacecraft, propulsion systems, and subsystems that exist on aerospace vehicles.
- Aircraft structures: Design of the physical configuration of the craft to withstand the forces encountered during flight. Aerospace engineering aims to keep structures lightweight.
- Materials science: Related to structures, aerospace engineering also studies the materials of which the aerospace structures are to be built. New materials with very specific properties are invented, or existing ones are modified to improve their performance.
- Solid mechanics: Closely related to material science is solid mechanics which deals with stress and strain analysis of the components of the vehicle. Nowadays there are several Finite Element programs such as MSC Patran/Nastran which aid engineers in the analytical process.
- Aeroelasticity: The interaction of aerodynamic forces and structural flexibility, potentially causing flutter, divergence, etc.

- Avionics: The design and programming of computer systems on board an aircraft or spacecraft and the simulation of systems.
- Risk and reliability: The study of risk and reliability assessment techniques and the mathematics involved in the quantitative methods.
- Noise control: The study of the mechanics of sound transfer.
- Flight test: Designing and executing flight test programs in order to gather and analyze performance and handling qualities data in order to determine if an aircraft meets its design and performance goals and certification requirements.

The basis of most of these components lies in theoretical mathematics, such as fluid dynamics for aerodynamics or the equations of motion for flight dynamics. In addition, there is a large, empirical component. Historically, this empirical component was derived from the testing of scale models and prototypes, either in wind tunnels or in the free atmosphere. More recently, advances in computing have enabled the use of computational fluid dynamics to simulate the behavior of fluid, reducing time and expense spent on wind-tunnel testing.

Additionally, aerospace engineering addresses the integration of all components that constitute an aerospace vehicle (subsystems including power, communications, thermal control, life support, and so forth) and its life cycle (design, temperature, pressure, radiation, velocity, life time).

## References


- Aerospace-products-manufacturers-and-markets, technology-aerospace-industry-225641: britannica.com, Retrieved 02 January, 2020
- Unmanned-aerial-vehicle, technology: britannica.com, Retrieved 09 June, 2020
- Flying-wing-aircraft, aircraft: militaryfactory.com, Retrieved 13 March, 2020
- Glider-aircraft, technology: britannica.com, Retrieved 19 February, 2020
- Rotorcraft: definitions.net, Retrieved 24 April, 2020
- What-is-a-helicopter: nasa.gov, Retrieved 02 August, 2020
- Aerospace-engineering: newworldencyclopedia.org, Retrieved 29 July, 2020

# 2

# Aerodynamics

Aerodynamics is the sub-field of fluid and gas dynamics that studies the interaction between the motion of air and solid objects. Subsonic and transonic flow, flight dynamics, supersonic and hypersonic speed, etc. are some of its focus areas. The topics elaborated in this chapter will help in gaining a better understanding of aerodynamics.

Aerodynamics is the study of how gases interact with moving bodies. Because the gas that we encounter most is air, aerodynamics is primarily concerned with the forces of drag and lift, which are caused by air passing over and around solid bodies. Engineers apply the principles of aerodynamics to the designs of many different things, including buildings, bridges and even soccer balls; however, of primary concern is the aerodynamics of aircraft and automobiles.

Aerodynamics comes into play in the study of flight and the science of building and operating an aircraft, which is called aeronautics. Aeronautical engineers use the fundamentals of aerodynamics to design aircraft that fly through the Earth's atmosphere.

## Aerodynamics of Aircraft

In order to overcome drag forces, an aircraft must generate thrust. This is accomplished with a motor-driven propeller or a jet engine. When the airplane is in level flight at a constant speed, the force of the thrust is just enough to counteract the aerodynamic drag.

Moving air can also generate forces in a different direction from the flow. The force that keeps an airplane from falling is called lift. Lift is generated by an aircraft wing. The path over a wing's curved top is longer than the path along the flat bottom of the wing. This causes the air to move faster over the top than it does along the bottom. With all other factors being equal, faster moving air has lower pressure than slower moving air, according to Bernoulli's principle, stated by Daniel Bernoulli, one of the most important pioneers in the field of fluid dynamics. This difference is what allows the slower moving air to push up against the bottom of the wing with greater force than the faster moving air is pushing down against the top of the wing. In level flight, this upward force is just enough to counteract the downward force caused by gravity.

Aerodynamic forces are also used to control an aircraft in flight. When the Wright brothers made their first flight in 1903, they needed a way to control their aircraft to climb, descend, bank and turn. They developed what is known as three-axis control for pitch, roll and yaw. Pitch (nose pointing up or down) is controlled by an elevator (the "flaps") on the back or trailing edge of the horizontal stabilizer in the tail section. Roll (tilting left or right) is controlled by ailerons (also flaps) on the trailing edges of the wings near the tips. Yaw (nose pointing left or right) is controlled by the rudder on the trailing edge of the vertical stabilizer in the tail section. These controls employ Newton's Third Law of Motion because they generate force by deflecting the airflow in the opposite direction of the desired movement. This force is also what allows aerobatic planes to fly upside down.

A pilot may also use flaps on the inboard section of the trailing edge of the wing during takeoff and landing. When in the downward position, flaps increase both lift and drag to allow the plane to fly slower without stalling. Some larger aircraft can also extend slats on the front or leading edges of the wings to increase lift at low speeds.

When the smooth airflow over a plane's wing is disrupted and this reduces the amount of lift, a stall can occur. According to the Federal Aviation Administration's Airplane Flying Handbook, "This is caused when the wing exceeds its critical angle of attack. This can occur at any airspeed, in any attitude, with any power setting." Typically, most stalls occur when an aircraft is moving too slowly with the nose at too high of an upward angle. The air no longer flows along the top surface but instead breaks away and forms turbulent swirls on top of the wing. This causes the plane to lose lift and start to fall, sometimes rather abruptly.

Another thing that can happen in an airplane is a spin. The Airplane Flying Handbook defines a spin as "an aggravated stall that results in what is termed 'autorotation' wherein the airplane follows a downward corkscrew path." This usually occurs in a slow turn when the slower inside wing stalls, and the outside wing is still generating lift. "Especially at low altitude, successful spin recovery may be difficult if not impossible in many aircraft," for the Canadian Owners and Pilots Association. One reason for this is the danger of going into a flat spin in which both wings and all of the control surfaces are stalled, and the aircraft falls like a maple tree seed.

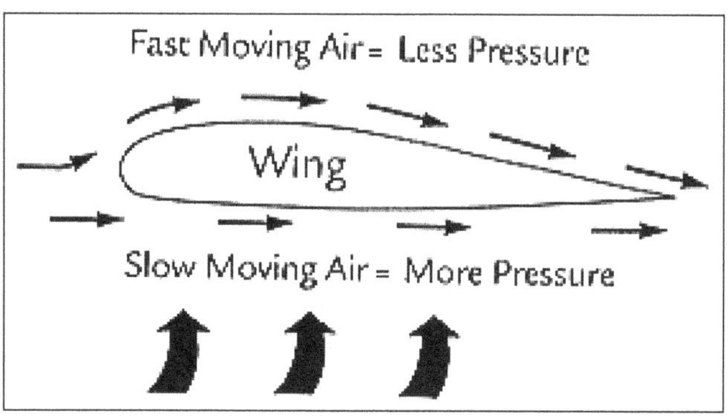


When a fluid is moving faster, it has lower pressure. This principle explains the lift created by an airplane's wing.

## SUBSONIC FLOW

Subsonic (or low-speed) aerodynamics describes fluid motion in flows which are much lower than the speed of sound everywhere in the flow. There are several branches of subsonic flow but one special case arises when the flow is inviscid, incompressible and irrotational. This case is called potential flow and allows the differential equations that describe the flow to be a simplified version of the equations of fluid dynamics, thus making available to the aerodynamicist a range of quick and easy solutions.

In solving a subsonic problem, one decision to be made by the aerodynamicist is whether to incorporate the effects of compressibility. Compressibility is a description of the amount of change of density in the flow. When the effects of compressibility on the solution are small, the assumption that density is constant may be made. The problem is then an incompressible low-speed aerodynamics problem. When the density is allowed to vary, the flow is called compressible. In air, compressibility effects are usually ignored when the Mach number in the flow does not exceed 0.3 (about 335 feet (102 m) per second or 228 miles (366 km) per hour at 60 °F (16 °C)). Above Mach 0.3, the problem flow should be described using compressible aerodynamics.

## TRANSONIC FLOW

Aerodynamic condensation evidences of supersonic expansion fans around a transonic F/A-18.

In aeronautics, transonic (or transsonic) flight is flying at or near the speed of sound 343 meters per second (1,235 km/h; 1,125 ft/s; 767 mph; 667 kn, at sea level under average conditions), relative to the air through which the vehicle is traveling. A typical convention used is to define transonic flight as speeds in the range of Mach 0.72 to 1.0 (965–1,235 km/h (600–767 mph) at sea level).

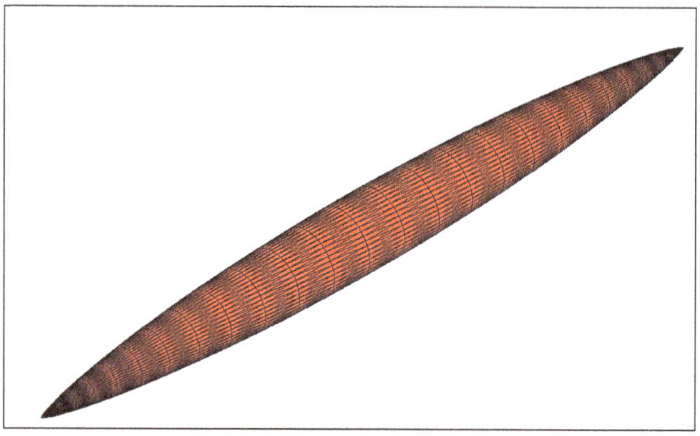

The Sears–Haack body presents a cross-sectional area variation that minimises wave drag.

This condition depends not only on the travel speed of the craft, but also on the temperature of the airflow in the vehicle's local environment. It is formally defined as the range of speeds between the critical Mach number, when some parts of the airflow over an air vehicle or airfoil are supersonic, and a higher speed, typically near Mach 1.2, when most of the airflow is supersonic. Between these speeds some of the airflow is supersonic, but a significant fraction is not.

Shock waves may appear as weak optical disturbances above airliners with supercritical wings.

Most modern jet powered aircraft are engineered to operate at transonic air speeds. Transonic airspeeds see a rapid increase in drag from about Mach 0.8, and it is the fuel costs of the drag that typically limits the airspeed. Attempts to reduce wave drag can be seen on all high-speed aircraft. Most notable is the use of swept wings, but another common form is a wasp-waist fuselage as a side effect of the Whitcomb area rule.

Severe instability can occur at transonic speeds. Shock waves can cause large-scale separation downstream, increasing drag and adding asymmetry and unsteadiness to the flow around the vehicle. Research has been done into weakening shock waves in transonic flight through the use of anti-shock bodies and supercritical airfoils.

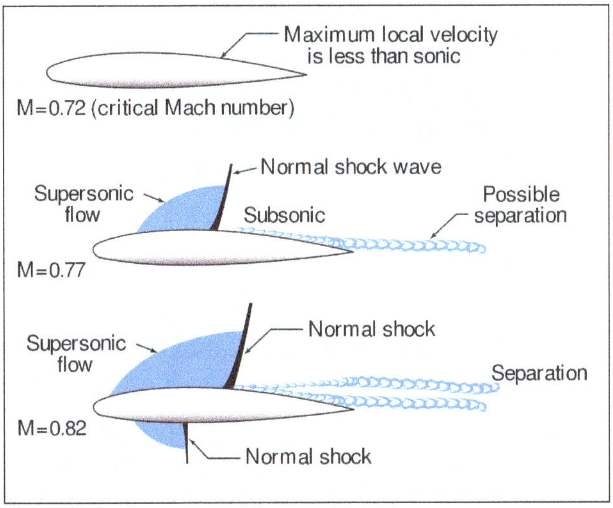


Transonic flow patterns on an airfoil showing flow patterns at and above critical Mach number.

Transonic speeds can also occur at the tips of rotor blades of helicopters and aircraft. This puts severe, unequal stresses on the rotor blade and may lead to accidents if it occurs. It is one of the limiting factors of the size of rotors and the forward speeds of helicopters (as this speed is added to the forward-sweeping [leading] side of the rotor, possibly causing localized transonics).

## Condensation Clouds

At transonic speeds supersonic expansion fans form intense low-pressure, low-temperature areas at various points around an aircraft. If the temperature drops below the dew point a visible cloud will form. These clouds remain with the aircraft as it travels. It is not necessary for the aircraft as a whole to reach supersonic speeds for these clouds to form. Typically, the tail of the aircraft will reach supersonic flight while the bow of the aircraft is still in subsonic flight. A bubble of supersonic expansion fans terminating by a wake shockwave surround the tail. As the aircraft continues to accelerate, the supersonic expansion fans will intensify and the wake shockwave will grow in size until infinity is reached, at which point the bow shockwave forms. This is Mach 1 and the Prandtl–Glauert singularity.

## Transonic Flows in Astronomy and Astrophysics

In astrophysics, wherever there is evidence of shocks (standing, propagating or oscillating), the flow close by must be transonic, as only supersonic flows form shocks. All black hole accretions are transonic. Many such flows also have shocks very close to the black holes.

The outflows or jets from young stellar objects or disks around black holes can also be transonic since they start subsonically and at a far distance they are invariably supersonic. Supernovae explosions are accompanied by supersonic flows and shock waves. Bow shocks formed in solar winds are a direct result of transonic winds from a star. It has been long thought that a bow shock was present around the heliosphere of our solar system. This was recently found not to be the case according to IBEX data.

# SUPERSONIC SPEED

Supersonic travel is a rate of travel of an object that exceeds the speed of sound (Mach 1). For objects traveling in dry air of a temperature of 20 °C (68 °F) at sea level, this speed is approximately 344 m/s, 1,125 ft/s, 768 mph, 667 knots, or 1,235 km/h. Speeds greater than five times the speed of sound (Mach 5) are often referred to as hypersonic. Flights during which only some parts of the air surrounding an object, such as the ends of rotor blades, reach supersonic speeds are called transonic. This occurs typically somewhere between Mach 0.8 and Mach 1.2.

Sounds are traveling vibrations in the form of pressure waves in an elastic medium. In gases, sound travels longitudinally at different speeds, mostly depending on the molecular mass and temperature of the gas, and pressure has little effect. Since air temperature and composition varies significantly with altitude, Mach numbers for aircraft may change despite a constant travel speed. In water at room temperature supersonic speed can be considered as any speed greater than 1,440 m/s (4,724 ft/s). In solids, sound waves can be polarized longitudinally or transversely and have even higher velocities.

Supersonic fracture is crack motion faster than the speed of sound in a brittle material.

## Supersonic Objects

British Airways Concorde in early BA livery at London-Heathrow Airport, in the early 1980s.

The tip of a bullwhip is thought to be the first man-made object to break the sound barrier, resulting in the telltale "crack" (actually a small sonic boom). The wave motion traveling through the bullwhip is what makes it capable of achieving supersonic speeds.

Most modern fighter aircraft are supersonic aircraft, but there have been supersonic passenger aircraft, namely Concorde and the Tupolev Tu-144. Both these passenger aircraft and some modern fighters are also capable of supercruise, a condition of sustained supersonic flight without the use of an afterburner. Due to its ability to supercruise for several hours and the relatively high frequency of flight over several decades, Concorde spent more time flying supersonically than all other aircraft combined by a considerable margin. Since Concorde's final retirement flight on November 26, 2003, there are no supersonic passenger aircraft left in service. Some large bombers, such as the Tupolev Tu-160 and Rockwell B-1 Lancer are also supersonic-capable.

Most modern firearm bullets are supersonic, with rifle projectiles often travelling at speeds approaching and in some cases well exceeding Mach 3.

Most spacecraft, most notably the Space Shuttle are supersonic at least during portions of their re-entry, though the effects on the spacecraft are reduced by low air densities. During ascent, launch vehicles generally avoid going supersonic below 30 km (~98,400 feet) to reduce air drag.

Note that the speed of sound decreases somewhat with altitude, due to lower temperatures found there (typically up to 25 km). At even higher altitudes the temperature starts increasing, with the corresponding increase in the speed of sound.

When an inflated balloon is burst, the torn pieces of latex contract at supersonic speed, which contributes to the sharp and loud popping noise.

## Supersonic Land Vehicles

To date, only one land vehicle has officially travelled at supersonic speed. It is ThrustSSC, driven by Andy Green, which holds the world land speed record, having achieved an average speed on its bi-directional run of 1,228 km/h (763 mph) in the Black Rock Desert on 15 October 1997.

The Bloodhound LSR project is planning an attempt on the record in 2020 at Hakskeen Pan in South Africa with a combination jet and hybrid rocket propelled car. The aim is to break the existing record, then make further attempts during which the team hope to reach speeds of up to 1,600 km/h (1,000 mph). The effort was originally run by Richard Noble who was the leader of the ThrustSSC project, however following funding issues in 2018, the team was bought by Ian Warhurst and renamed Bloodhound LSR. The new project retains many of the original Bloodhound SSC engineering staff, and Andy Green is still the driver for record attempt, with high speed trials expected to start in October 2019.

## Supersonic Flight

Supersonic aerodynamics is simpler than subsonic aerodynamics because the airsheets at different points along the plane often cannot affect each other. Supersonic jets and rocket vehicles require several times greater thrust to push through the extra aerodynamic drag experienced within the transonic region (around Mach 0.85–1.2). At these speeds aerospace engineers can gently guide air around the fuselage of the aircraft without producing new shock waves, but any change in cross area farther down the vehicle leads to shock waves along the body. Designers use the Supersonic area rule and the Whitcomb area rule to minimize sudden changes in size.

The sound source has now broken through the sound speed barrier, and is traveling at 1.4 times the speed of sound, c (Mach 1.4). Because the source is moving faster than the sound waves it creates, it actually leads the advancing wavefront. The sound source will pass by a stationary observer before the observer actually hears the sound it creates.

However, in practical applications, a supersonic aircraft must operate stably in both subsonic and supersonic profiles, hence aerodynamic design is more complex.

One problem with sustained supersonic flight is the generation of heat in flight. At high speeds aerodynamic heating can occur, so an aircraft must be designed to operate and function under very high temperatures. Duralumin, the traditional aircraft material, starts to lose strength and go into plastic deformation at relatively low temperatures, and is unsuitable for continuous use at speeds above Mach 2.2 to 2.4. Materials such as titanium and stainless steel allow operations at much higher temperatures. For example, the Lockheed SR-71 Blackbird jet could fly continuously at Mach 3.1 which could lead to temperatures on some parts of the aircraft getting above 315 °C (600 °F).

Another area of concern for sustained high-speed flight is engine operation. Jet engines create thrust by increasing the temperature of the air they ingest, and as the aircraft speeds up, friction and compression heat this air before it reaches the engines. The maximum allowable temperature of the exhaust is determined by the materials in the turbine at the rear of the engine, so as the aircraft speeds up, the difference in intake and exhaust temperature that the engine can create decreases, and the thrust along with it. Air cooling the turbine area to allow operations at higher temperatures was a key solution, one that continued to improve through the 1950s and on to this day.

Intake design was also a major issue. Normal jet engines can only ingest subsonic air, so for supersonic operation the air must be slowed down. Ramps or cones in the intake are used to create shock waves that slow the airflow before it reaches the engine. Doing so removes energy from the airflow, causing drag. The key to reducing this drag is to use multiple small oblique shock waves, but this was difficult because the angle they make inside the intake changes with Mach number. In order to efficiently operate across a range of speeds, the shock waves have to be "tuned."

An aircraft able to operate for extended periods at supersonic speeds has a potential range advantage over a similar design operating subsonically. Most of the drag an aircraft sees while speeding up to supersonic speeds occurs just below the speed of sound, due to an aerodynamic effect known as wave drag. An aircraft that can accelerate past this speed sees a significant drag decrease, and can fly supersonically with improved fuel economy. However, due to the way lift is generated supersonically, the lift-to-drag ratio of the aircraft as a whole drops, leading to lower range, offsetting or overturning this advantage.

The key to having low supersonic drag is to properly shape the overall aircraft to be long and thin, and close to a "perfect" shape, the von Karman ogive or Sears-Haack body. This has led to almost every supersonic cruising aircraft looking very similar to every other, with a very long and slender fuselage and large delta wings, cf. SR-71, Concorde, etc. Although not ideal for passenger aircraft, this shaping is quite adaptable for bomber use.

## HYPERSONIC SPEED

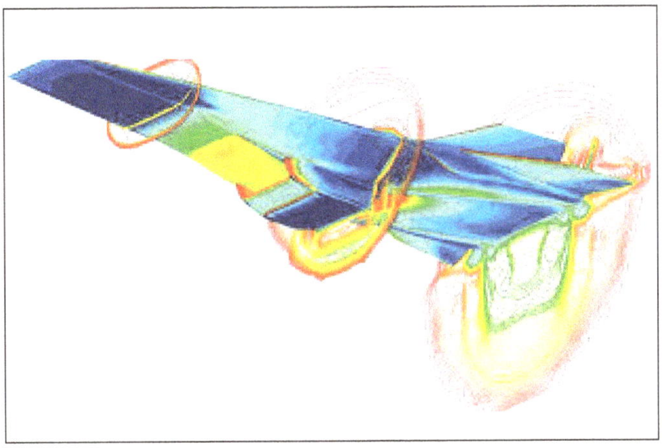

CFD image of the NASA X-43A at Mach 7.

In aerodynamics, a hypersonic speed is one that greatly exceeds the speed of sound, often stated as starting at speeds of Mach 5 and above.

The precise Mach number at which a craft can be said to be flying at hypersonic speed varies, since individual physical changes in the airflow (like molecular dissociation and ionization) occur at different speeds; these effects collectively become important around Mach 5-10. The hypersonic regime can also be alternatively defined as speeds where specific heat capacity changes with the temperature of the flow as kinetic energy of the moving object is converted into heat.

### Characteristics of Flow

While the definition of hypersonic flow can be quite vague and is generally debatable (especially due to the absence of discontinuity between supersonic and hypersonic flows), a hypersonic flow may be characterized by certain physical phenomena that can no longer be analytically discounted as in supersonic flow. The peculiarity in hypersonic flows are as follows:

- Shock layer.
- Aerodynamic heating.
- Entropy layer.
- Real gas effects.
- Low density effects.
- Independence of aerodynamic coefficients with Mach number.

## Small Shock Stand-off Distance

As a body's Mach number increases, the density behind a bow shock generated by the body also increases, which corresponds to a decrease in volume behind the shock due to conservation of mass. Consequently, the distance between the bow shock and the body decreases at higher Mach numbers.

## Entropy Layer

As Mach numbers increase, the entropy change across the shock also increases, which results in a strong entropy gradient and highly vortical flow that mixes with the boundary layer.

## Viscous Interaction

A portion of the large kinetic energy associated with flow at high Mach numbers transforms into internal energy in the fluid due to viscous effects. The increase in internal energy is realized as an increase in temperature. Since the pressure gradient normal to the flow within a boundary layer is approximately zero for low to moderate hypersonic Mach numbers, the increase of temperature through the boundary layer coincides with a decrease in density. This causes the bottom of the boundary layer to expand, so that the boundary layer over the body grows thicker and can often merge with the shock wave near the body leading edge.

## High-temperature Flow

High temperatures due to a manifestation of viscous dissipation cause non-equilibrium chemical flow properties such as vibrational excitation and dissociation and ionization of molecules resulting in convective and radiative heat-flux.

## Classification of Mach Regimes

Although "subsonic" and "supersonic" usually refer to speeds below and above the local speed of sound respectively, aerodynamicists often use these terms to refer to particular ranges of Mach values. This occurs because a "transonic regime" exists around M=1 where approximations of the Navier–Stokes equations used for subsonic design no longer apply, partly because the flow locally exceeds M=1 even when the freestream Mach number is below this value.

The "supersonic regime" usually refers to the set of Mach numbers for which linearised theory may be used; for example, where the (air) flow is not chemically reacting and where heat transfer between air and vehicle may be reasonably neglected in calculations. Generally, NASA defines "high" hypersonic as any Mach number from 10 to 25, and re-entry speeds as anything greater than Mach 25. Among the aircraft operating in this regime are the Space Shuttle and (theoretically) various developing spaceplanes.

## Similarity Parameters

The categorization of airflow relies on a number of similarity parameters, which allow the simplification of a nearly infinite number of test cases into groups of similarity. For transonic and compressible flow, the Mach and Reynolds numbers alone allow good categorization of many flow cases.

Hypersonic flows, however, require other similarity parameters. First, the analytic equations for the oblique shock angle become nearly independent of Mach number at high (~>10) Mach numbers. Second, the formation of strong shocks around aerodynamic bodies means that the freestream Reynolds number is less useful as an estimate of the behavior of the boundary layer over a body (although it is still important). Finally, the increased temperature of hypersonic flows mean that real gas effects become important. For this reason, research in hypersonics is often referred to as aerothermodynamics, rather than aerodynamics.

The introduction of real gas effects means that more variables are required to describe the full state of a gas. Whereas a stationary gas can be described by three variables (pressure, temperature, adiabatic index), and a moving gas by four (flow velocity), a hot gas in chemical equilibrium also requires state equations for the chemical components of the gas, and a gas in nonequilibrium solves those state equations using time as an extra variable. This means that for a nonequilibrium flow, something between 10 and 100 variables may be required to describe the state of the gas at any given time. Additionally, rarefied hypersonic flows (usually defined as those with a Knudsen number above 0.1) do not follow the Navier–Stokes equations.

Hypersonic flows are typically categorized by their total energy, expressed as total enthalpy (MJ/kg), total pressure (kPa-MPa), stagnation pressure (kPa-MPa), stagnation temperature (K), or flow velocity (km/s).

Wallace D. Hayes developed a similarity parameter, similar to the Whitcomb area rule, which allowed similar configurations to be compared.

## Regimes

Hypersonic flow can be approximately separated into a number of regimes. The selection of these regimes is rough, due to the blurring of the boundaries where a particular effect can be found.

## Perfect Gas

In this regime, the gas can be regarded as an ideal gas. Flow in this regime is still Mach number dependent. Simulations start to depend on the use of a constant-temperature wall, rather than the adiabatic wall typically used at lower speeds. The lower border of this region is around Mach 5, where ramjets become inefficient, and the upper border around Mach 10-12.

## Two-temperature Ideal Gas

This is a subset of the perfect gas regime, where the gas can be considered chemically perfect, but the rotational and vibrational temperatures of the gas must be considered separately, leading to two temperature models.

## Dissociated Gas

In this regime, diatomic or polyatomic gases (the gases found in most atmospheres) begin to dissociate as they come into contact with the bow shock generated by the body. Surface catalysis plays a role in the calculation of surface heating, meaning that the type of surface material also has an effect on the flow. The lower border of this regime is where any component of a gas mixture first begins to dissociate in the stagnation point of a flow (which for nitrogen is around 2000 K). At the upper border of this regime, the effects of ionization start to have an effect on the flow.

## Ionized Gas

In this regime the ionized electron population of the stagnated flow becomes significant, and the electrons must be modeled separately. Often the electron temperature is handled separately from the temperature of the remaining gas components. This region occurs for freestream flow velocities around 10–12 km/s. Gases in this region are modeled as non-radiating plasmas.

## Radiation-dominated Regime

Above around 12 km/s, the heat transfer to a vehicle changes from being conductively dominated to radiatively dominated. The modeling of gases in this regime is split into two classes:

- Optically thin: Where the gas does not re-absorb radiation emitted from other parts of the gas.

- Optically thick: Where the radiation must be considered a separate source of energy.

The modeling of optically thick gases is extremely difficult, since, due to the calculation of the radiation at each point, the computation load theoretically expands exponentially as the number of points considered increases.

# AERODYNAMIC HEATING

Aerodynamic heating is the heating of a solid body produced by its high-speed passage through air (or by the passage of air past a test object in a wind tunnel), whereby its kinetic energy is converted to heat by adiabatic heating, and (less significantly) by skin friction on the surface of the object at a rate that depends on the viscosity and speed of the air. In science and engineering, it is most frequently a concern regarding meteors, atmospheric reentry of spacecraft, and the design of high-speed aircraft.

## Physics

At high speeds through the air, the object's kinetic energy is converted to heat through compression and friction. At lower speed, the object will lose heat to the air through which it is passing, if the air is cooler. The combined temperature effect of heat from the air and from passage through it is called the stagnation temperature; the actual temperature is called the recovery temperature. These viscous dissipative effects to neighboring sub-layers make the boundary layer slow down via a non-isentropic process. Heat then conducts into the surface material from the higher temperature air. The result is an increase in the temperature of the material and a loss of energy from the flow. The forced convection ensures that other material replenishes the gases that have cooled to continue the process.

The stagnation and the recovery temperature of a flow increases with the speed of the flow and are greater at high speeds. The total thermal loading of the object is a function of both the recovery temperature and the mass flow rate of the flow. Aerodynamic heating is greatest at high speed and in the lower atmosphere where the density is greater. In addition to the convective process, there is also Thermal radiation from the flow to the body and vice versa with the net direction set by the relative temperature of each.

Aerodynamic heating increases with the speed of the vehicle. Its effects are minimal at subsonic speeds but at supersonic speeds beyond about M2.2 it dictates the design/materials of the vehicle structure and internal systems. The heating effects are greatest at leading edges but the whole vehicle heats up to a stabilized temperature if it remains at speed. Aerodynamic heating is dealt with by the use of high temperature alloys for metals, the addition of insulation of the exterior of the vehicle, or the use of ablative material.

## Aircraft

Aerodynamic heating is a concern for supersonic and hypersonic aircraft.

One of the main concerns caused by aerodynamic heating arises in the design of the wing. When the structure of an aircraft wing is designed, there are two main considerations that must be accounted for when this aircraft is to fly at subsonic speeds: minimizing weight and maximizing strength. Aerodynamic heating, which occurs at supersonic and hypersonic aircraft speeds, adds an additional consideration in wing structure analysis. In an idealized wing structure, a wing is made up of spars, stringers, and skin segments. In a wing that normally experiences subsonic speeds, there must be a sufficient number of stringers to withstand the axial and bending stresses induced by the lift force acting on the wing. In addition, the distance between the stringers must be small enough that the skin panels do not buckle, and the panels must be thick enough to withstand the shear stress and shear flow present in the panels due to the lifting force on the wing. However, the weight of the wing must be made as small as possible, so the choice of material for the stringers and the skin is an important factor.

At supersonic airspeeds, aerodynamic heating adds another element to this structural analysis. At normal speeds, spars and stringers experience a load called Delta P, which is a function of the lift force, first and second moments of inertia, and length of the spar. When there are more spars and stringers, the Delta P in each member is reduced, and the area of the stringer can be reduced to meet critical stress requirements. However, the increase in temperature caused by energy flowing from the air (heated by skin friction at these high speeds) adds another load factor, called a thermal load, to the spars. This thermal load increases the net force felt by the stringers, and thus the area of the stringers must be increased in order for the critical stress requirement to be met.

Another issue that aerodynamic heating causes for aircraft design is the effect of high temperatures on common material properties. Common materials used in aircraft wing design, such as aluminum and steel, experience a decrease in strength as temperatures get extremely high. The Young's Modulus of the material, defined as the ratio between stress and strain experienced by the material, decreases as the temperature increases. Young's Modulus is critical in the selection of materials for wing, as a higher value lets the material resist the yield and shear stress caused by the lift and thermal loads. This is because Young's Modulus is an important factor in the equations for calculating the critical buckling load for axial members and the critical buckling shear stress for skin panels. If the Young's Modulus of the material decreases at high temperatures caused by aerodynamic heating, then the wing design will call for larger spars and thicker skin segments in order to account for this decrease in strength as the aircraft goes supersonic. There are some materials that retain their strength at the high temperatures that aerodynamic heating induces. For example, Inconel X was used for the wing skins of the X-15, a North American aircraft that flew at hypersonic speeds in 1958. Titanium is another high-strength material, even at high temperatures, and is often used for wing frames of supersonic aircraft. The SR-71 used titanium skin panels painted black to reduce the temperature and corrugated to accommodate expansion. Another important design concept for early supersonic aircraft wings was using a small thickness-to-chord ratio, so that the speed of the flow over the airfoil does not increase too much from the free stream speed. As the flow is already supersonic, increasing the speed even more would not be beneficial for the wing structure.

Reducing the thickness of the wing brings the top and bottom stringers closer together, reducing the total moment of inertia of the structure. This increases is axial load in the stringers, and thus the area, and weight, of the stringers must be increased. Some designs for hypersonic missiles have used liquid cooling of the leading edges (usually the fuel en route to the engine). The Sprint missile's heat shield needed several design iterations for Mach 10 temperatures.

### Reentry Vehicles

Heating caused by the very high reentry speeds (greater than Mach 20) is sufficient to destroy the vehicle unless special techniques are used. The early space capsules such as used on Mercury, Gemini, and Apollo were given blunt shapes to produce a stand-off bow shock, allowing most of the heat to dissipate into the surrounding air. Additionally, these vehicles had ablative material that sublimates into a gas at high temperature. The act of sublimation absorbs the thermal energy from the aerodynamic heating and erodes the material away as opposed to heating the capsule. The surface of the heat shield for the Mercury spacecraft had a coating of aluminum with glassfiber in many layers. As the temperature rose to 1,100 °C (1,400 K) the layers would evaporate and take the heat with it. The spacecraft would become hot but not harmfully so. The Space Shuttle used insulating tiles on its lower surface to absorb and radiate heat while preventing conduction to the aluminium airframe. The damage to the heat shield during liftoff of Space Shuttle *Columbia* contributed to its destruction upon reentry.

## ANGLE OF ATTACK

Angle of Attack is used to define the angle between the wing chord line and the flight path. This is not to be confused with the relation of the aircraft to the Earth's surface. This is called the attitude and is seldom, if ever, the same as the angle of attack.

When the angle is small, the aircraft is said to be at a low angle of attack. When the angle is large, the aircraft is said to be at a high angle of attack. During landing, an airplane may have a level attitude, but a high angle of attack, because the flight path is downward and the approaching wind is parallel to the flight path. During climb, an airplane can be in a nose-high attitude, but at a low angle of attack.

Two variables can change the amount of lift generated by a wing in a given configuration.

- The speed of air flowing over the airfoil.
- The angle of attack.

An increase in speed or the angle of attack will increase both lift and drag. In level flight, lift must equal the weight of the aircraft. If an aircraft weighs 2,000 lbs., the wing must generate 2,000 lbs. of lift. The speed and angle of attack are interchangeable to a point—therefore, for every airspeed, there is a corresponding angle of attack that will produce the same amount of lift.

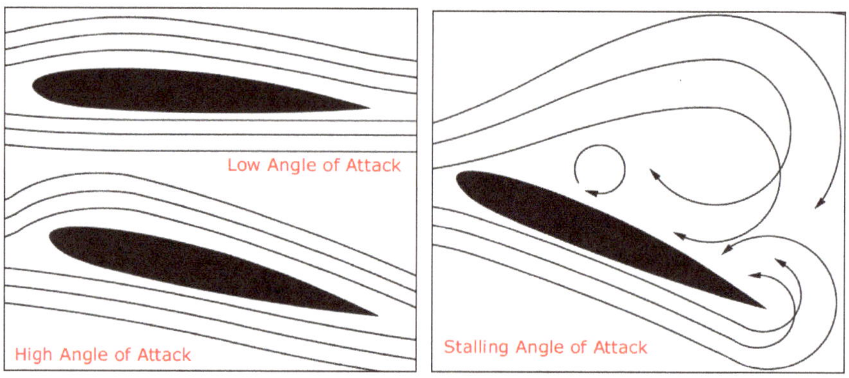


In order for a wing to produce lift, the air flowing past an aircraft, must be aligned to the airfoil in order to provide a smooth airflow. As a wing increases its angle of attack, airflow can no longer flow smoothly over the wing and eddies or burbles will form, causing the wing to approach its stall speed. When a wing finally stalls, it will no longer produce lift and with weight unopposed by lift, the aircraft will drop towards the ground. With sufficient altitude, stall recovery can be obtained by decreasing the angle of attack.

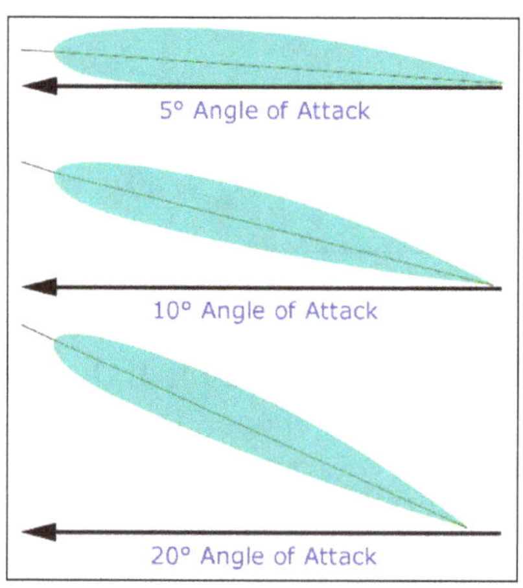


Angle of Attack is the angle between the wing chord line and the flight path.

An airplane can be stalled at any attitude and at any airspeed such as pulling out of a dive too abruptly or if the airplane is in a steep turn at a high angle of attack, even though the airspeed is high. During dives or turns, centrifugal force will increase load factors and if excessive, this will cause the wing to stall.

The angle of attack should not be confused with the angle of incidence. The angle of incidence is the angle formed by the wing chord line and the aircraft longitudinal axis.

# FLIGHT DYNAMICS

Flight dynamics is the study of the performance, stability, and control of vehicles flying through the air or in outer space. It is concerned with how forces acting on the vehicle influence its speed and attitude with respect to time.

For a fixed-wing aircraft, its changing orientation with respect to the local air flow is represented by two critical angles, the angle of attack of the wing ("alpha") and the angle of attack of the vertical tail, known as the sideslip angle ("beta"). A sideslip angle will arise if an aircraft yaws about its centre of gravity and if the aircraft sideslips bodily, i.e. the centre of gravity moves sideways. These angles are important because they are the principal source of changes in the aerodynamic forces and moments applied to the aircraft.

Spacecraft flight dynamics involve three forces: propulsive (rocket engine), gravitational, and lift and drag (when traveling through the earth's or any other atmosphere). Because aerodynamic forces involved with spacecraft flight are very small, this leaves gravity as the dominant force.

## Aircraft

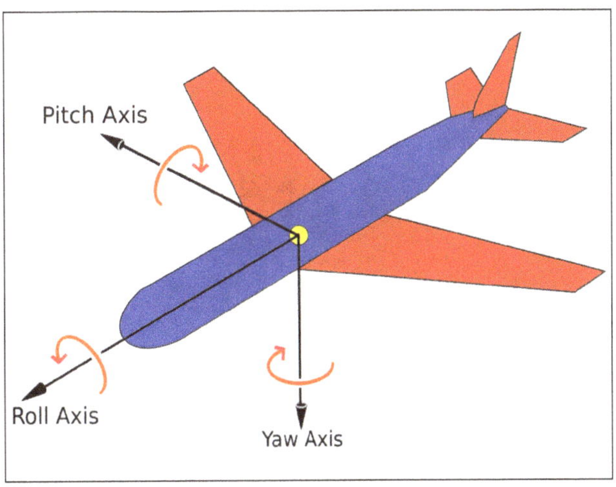


Axes to control the attitude of a plane.

Flight dynamics is the science of air-vehicle orientation and control in three dimensions. The critical flight dynamics parameters are the angles of rotation with respect to the three aircraft's principal axes about its center of gravity, known as roll, pitch and yaw.

Aircraft engineers develop control systems for a vehicle's orientation (attitude) about its center of gravity. The control systems include actuators, which exert forces in various directions, and generate rotational forces or moments about the center of gravity of the aircraft, and thus rotate the aircraft in pitch, roll, or yaw. For example, a pitching moment is a vertical force applied at a distance forward or aft from the center of gravity of the aircraft, causing the aircraft to pitch up or down.

Roll, pitch and yaw refer, in this context, to rotations about the respective axes starting from a defined equilibrium state. The equilibrium roll angle is known as wings level or zero bank angle, equivalent to a level heeling angle on a ship. Yaw is known as "heading".

A fixed-wing aircraft increases or decreases the lift generated by the wings when it pitches nose up or down by increasing or decreasing the angle of attack (AOA). The roll angle is also known as bank angle on a fixed-wing aircraft, which usually "banks" to change the horizontal direction of flight. An aircraft is streamlined from nose to tail to reduce drag making it advantageous to keep the sideslip angle near zero, though aircraft are deliberately "side-slipped" when landing in a cross-wind, as explained in slip (aerodynamics).

## Spacecraft and Satellites

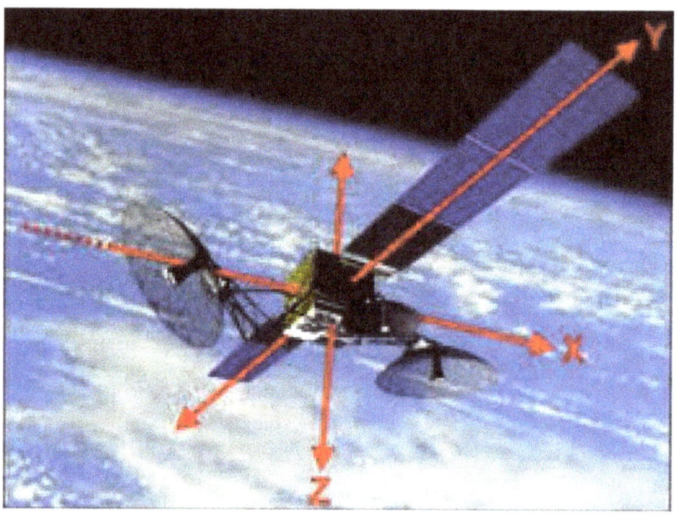


Axes to control the attitude of a satellite.

The forces acting on spacecraft are of three types: propulsive force (usually provided by the vehicle's engine thrust); gravitational force exerted by the Earth and other celestial bodies; and aerodynamic lift and drag (when flying in the atmosphere of the Earth or another body, such as Mars or Venus). The vehicle's attitude must be taken into account because of its effect on the aerodynamic and propulsive forces. There are other reasons, unrelated to flight dynamics, for controlling the vehicle's attitude in non-powered flight (e.g., thermal control, solar power generation, communications, or astronomical observation).

The flight dynamics of spacecraft differ from those of aircraft in that the aerodynamic forces are of very small, or vanishingly small effect for most of the vehicle's flight, and cannot be used for attitude control during that time. Also, most of a spacecraft's flight time is usually unpowered, leaving gravity as the dominant force.

## Flight Dynamics (Fixed-wing Aircraft)

Flight dynamics is the science of air vehicle orientation and control in three dimensions. The three critical flight dynamics parameters are the angles of rotation in three dimensions about the vehicle's center of gravity (cg), known as *pitch*, *roll* and *yaw*.

Control systems adjust the orientation of a vehicle about its cg. A control system includes control surfaces which, when deflected, generate a moment (or couple from ailerons) about the cg which rotates the aircraft in pitch, roll, and yaw. For example, a pitching moment comes from a force applied at a distance forward or aft of the cg, causing the aircraft to pitch up or down.

Roll, pitch and yaw refer to rotations about the respective axes starting from a defined steady flight equilibrium state. The equilibrium roll angle is known as wings level or zero bank angle.

The most common aeronautical convention defines roll as acting about the longitudinal axis, positive with the starboard (right) wing down. Yaw is about the vertical body axis, positive with the nose to starboard. Pitch is about an axis perpendicular to the longitudinal plane of symmetry, positive nose up.

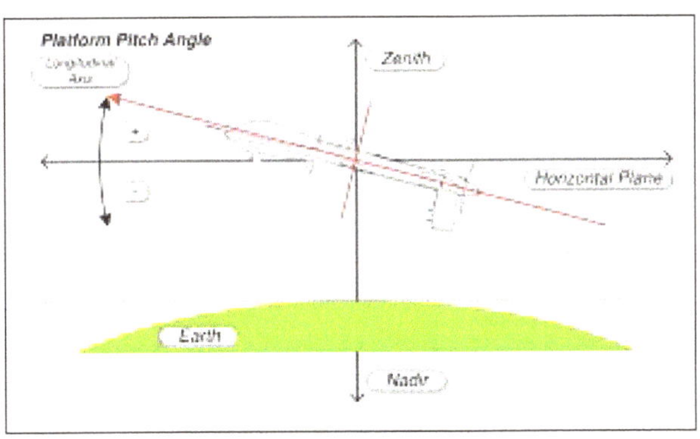


Pitch angle definition.

A fixed-wing aircraft increases or decreases the lift generated by the wings when it pitches nose up or down by increasing or decreasing the angle of attack (AOA). The roll angle is also known as bank angle on a fixed-wing aircraft, which usually "banks" to change the horizontal direction of flight. An aircraft is streamlined from nose to tail to reduce drag making it advantageous to keep the sideslip angle near zero, though an aircraft may be deliberately "sideslipped" to increase drag and descent rate during landing, to keep aircraft heading same as runway heading during cross-wind landings and during flight with asymmetric power.

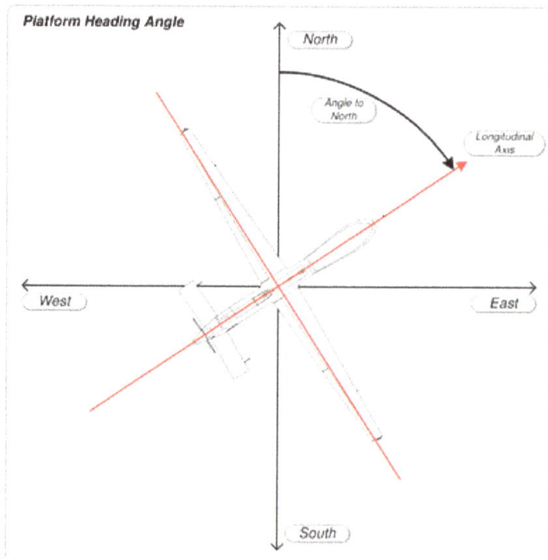


Yaw or heading angle definition.

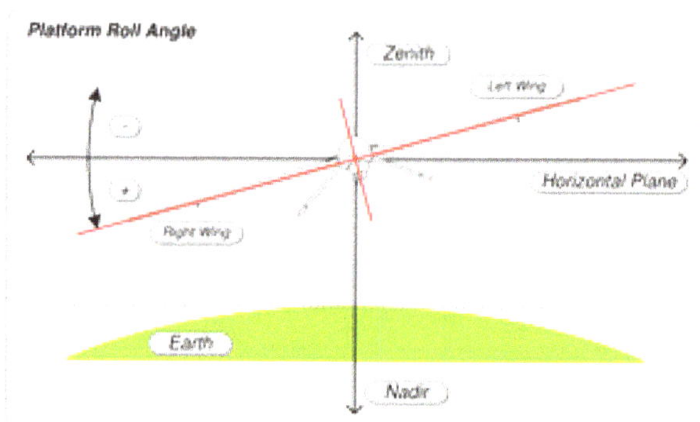


Roll angle definition.

## Reference Frames

Three right-handed, Cartesian coordinate systems see frequent use in flight dynamics. The first coordinate system has an origin fixed in the reference frame of the Earth.

- Earth frame:
    - Origin: Arbitrary, fixed relative to the surface of the Earth.

- $x_E$ axis: Positive in the direction of north.
- $y_E$ axis: Positive in the direction of east.
- $z_E$ axis: Positive towards the center of the Earth.

In many flight dynamics applications, the Earth frame is assumed to be inertial with a flat $x_E,y_E$-plane, though the Earth frame can also be considered a spherical coordinate system with origin at the center of the Earth.

The other two reference frames are body-fixed, with origins moving along with the aircraft, typically at the center of gravity. For an aircraft that is symmetric from right-to-left, the frames can be defined as:

- Body frame:
  - Origin: Airplane center of gravity.
  - $x_b$ axis: Positive out the nose of the aircraft in the plane of symmetry of the aircraft.
  - $z_b$ axis: Perpendicular to the $x_b$ axis, in the plane of symmetry of the aircraft, positive below the aircraft.
  - $y_b$ axis: Perpendicular to the $x_b$, $z_b$-plane, positive determined by the right-hand rule (generally, positive out the right wing).
- Wind frame:
  - Origin: Airplane center of gravity.
  - $x_w$ axis: Positive in the direction of the velocity vector of the aircraft relative to the air.
  - $z_w$ axis: Perpendicular to the $x_w$ axis, in the plane of symmetry of the aircraft, positive below the aircraft.
  - $y_w$ axis: Perpendicular to the $x_w$, $z_w$-plane, positive determined by the right hand rule (generally, positive to the right).

Asymmetric aircraft have analogous body-fixed frames, but different conventions must be used to choose the precise directions of the $x$ and $z$ axes.

The Earth frame is a convenient frame to express aircraft translational and rotational kinematics. The Earth frame is also useful in that, under certain assumptions, it can be approximated as inertial. Additionally, one force acting on the aircraft, weight, is fixed in the $+z_E$ direction.

The body frame is often of interest because the origin and the axes remain fixed relative to the aircraft. This means that the relative orientation of the Earth and body frames describes the aircraft attitude. Also, the direction of the force of thrust is generally fixed in the body frame, though some aircraft can vary this direction, for example by thrust vectoring.

The wind frame is a convenient frame to express the aerodynamic forces and moments acting on an aircraft. In particular, the net aerodynamic force can be divided into components along the wind frame axes, with the drag force in the $-x_w$ direction and the lift force in the $-z_w$ direction.

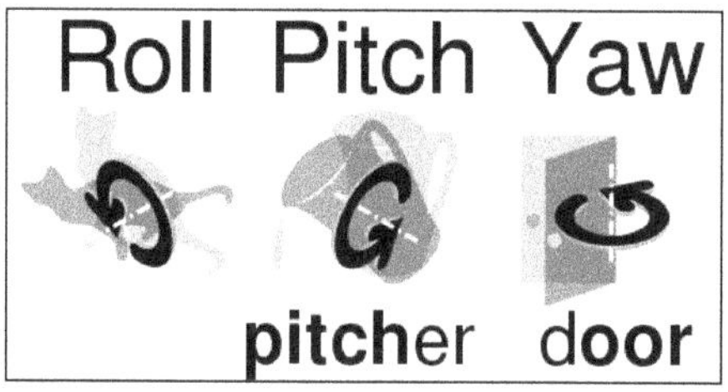

Mnemonics to remember angle names.

In addition to defining the reference frames, the relative orientation of the reference frames can be determined. The relative orientation can be expressed in a variety of forms, including:

- Direction cosine or rotation matrices.
- Euler angles.
- Quaternions.

The various Euler angles relating the three reference frames are important to flight dynamics. Many Euler angle conventions exist, but all of the rotation sequences presented below use the *z-y'-x"* convention. This convention corresponds to a type of Tait-Bryan angles, which are commonly referred to as Euler angles. This convention is described in detail below for the roll, pitch, and yaw Euler angles that describe the body frame orientation relative to the Earth frame. The other sets of Euler angles are described below by analogy.

To transform from the Earth frame to the body frame using Euler angles, the following rotations are done in the order prescribed. First, rotate the Earth frame axes $x_E$ and $y_E$ around the $z_E$ axis by the yaw angle $\psi$. This results in an intermediate reference frame with axes denoted $x',y',z'$, where $z'=z_E$. Second, rotate the $x'$ and $z'$ axes around the $y'$ axis by the pitch angle $\theta$. This results in another intermediate reference frame with axes denoted $x'',y'',z''$, where $y''=y'$. Finally, rotate the $y''$ and $z''$ axes around the $x''$ axis by the roll angle $\varphi$. The reference frame that results after the three rotations is the body frame.

Based on the rotations and axes conventions above, the yaw angle $\psi$ is the angle between north and the projection of the aircraft longitudinal axis onto the horizontal plane, the pitch angle $\theta$ is the angle between the aircraft longitudinal axis and horizontal, and the roll angle $\varphi$ represents a rotation around the aircraft longitudinal axis after rotating by yaw and pitch.

To transform from the Earth frame to the wind frame, the three Euler angles are the bank angle $\mu$, the flight path angle $\gamma$, and the heading angle $\sigma$. When performing the rotations described above to obtain the wind frame from the Earth frame, $(\mu,\gamma,\sigma)$ are analogous to $(\varphi, \theta,\psi)$, respectively. The heading angle $\sigma$ is the angle between north and the horizontal component of the velocity vector, which describes which direction the aircraft is moving relative to cardinal directions. The flight path angle $\gamma$ is the angle between horizontal and the velocity vector, which describes whether the aircraft is climbing or descending. The bank angle $\mu$ represents a rotation of the lift force around the velocity vector, which may indicate whether the airplane is turning.

To transform from the wind frame to the body frame, the two Euler angles are the angle of attack $\alpha$ and the sideslip angle $\beta$. When performing the rotations described earlier to obtain the body frame from the wind frame, $(\alpha, \beta)$ are analogous to $(\theta,\psi)$, respectively; the angle analogous to $\varphi$ in this transformation is always zero. The sideslip angle $\beta$ is the angle between the velocity vector and the projection of the aircraft longitudinal axis onto the $x_w,y_w$-plane, which describes whether there is a lateral component to the aircraft velocity, also known as sideslip. The angle of attack $\alpha$ is the angle between the $x_w,y_w$-plane and the aircraft longitudinal axis and, among other things, is an important variable in determining the magnitude of the force of lift.

## Design Cases

In analyzing the stability of an aircraft, it is usual to consider perturbations about a nominal steady flight state. So the analysis would be applied, for example, assuming:

- Straight and level flight.
- Turn at constant speed.
- Approach and landing.
- Takeoff.

The speed, height and trim angle of attack are different for each flight condition, in addition, the aircraft will be configured differently, e.g. at low speed flaps may be deployed and the undercarriage may be down.

Except for asymmetric designs (or symmetric designs at significant sideslip), the longitudinal equations of motion (involving pitch and lift forces) may be treated independently of the lateral motion (involving roll and yaw).

The following considers perturbations about a nominal straight and level flight path.

To keep the analysis (relatively) simple, the control surfaces are assumed fixed throughout the motion, this is stick-fixed stability. Stick-free analysis requires the further complication of taking the motion of the control surfaces into account.

Furthermore, the flight is assumed to take place in still air, and the aircraft is treated as a rigid body.

## Forces of Flight

Three forces act on an aircraft in flight: weight, thrust, and the aerodynamic force.

## Aerodynamic Force

### Components of the Aerodynamic Force

The expression to calculate the aerodynamic force is:

$$\mathbf{F}_A = \int_\Sigma (-\Delta p \mathbf{n} + \mathbf{f}) d\sigma$$

Where:

$\Delta p \equiv$ Difference between static pressure and free current pressure.

$\mathbf{n} \equiv$ outer normal vector of the element of area.

$\mathbf{f} \equiv$ tangential stress vector practised by the air on the body.

$\Sigma \equiv$ adequate reference surface.

Projected on wind axes we obtain:

$$\mathbf{F}_A = -(\mathbf{i}_w D + \mathbf{j}_w Q + \mathbf{k}_w L)$$

Where:

$D \equiv$ Drag

$Q \equiv$ Lateral force

$L \equiv$ Lift

### Aerodynamic Coefficients

Dynamic pressure of the free current $\equiv q = \frac{1}{2}\rho V^2$

Proper reference surface (wing surface, in case of planes) $\equiv S$

Pressure coefficient $\equiv C_p = \frac{p - p_\infty}{q}$

Friction coefficient $\equiv C_f = \frac{f}{q}$

Drag coefficient $\equiv C_d = \dfrac{D}{qS} = -\dfrac{1}{S}\int_{\Sigma}[(-C_p)\mathbf{n}\bullet\mathbf{i}_w + C_f\mathbf{t}\bullet\mathbf{i}_w]d\sigma$

Lateral force coefficient $\equiv C_Q = \dfrac{Q}{qS} = -\dfrac{1}{S}\int_{\Sigma}[(-C_p)\mathbf{n}\bullet\mathbf{j}_w + C_f\mathbf{t}\bullet\mathbf{j}_w]d\sigma$

Lift coefficient $\equiv C_L = \dfrac{L}{qS} = -\dfrac{1}{S}\int_{\Sigma}[(-C_p)\mathbf{n}\bullet\mathbf{k}_w + C_f\mathbf{t}\bullet\mathbf{k}_w]d\sigma$

It is necessary to know $C_p$ and $C_f$ in every point on the considered surface.

## Dimensionless Parameters and Aerodynamic Regimes

In absence of thermal effects, there are three remarkable dimensionless numbers:

- Compressibility of the flow:

  Mach number $\equiv M = \dfrac{V}{a}$.

- Viscosity of the flow:

  Reynolds number $\equiv Re = \dfrac{\rho V l}{\mu}$.

- Rarefaction of the flow:

  Knudsen number $\equiv Kn = \dfrac{\lambda}{l}$.

Where:

$a = \sqrt{kR\theta} \equiv$ speed of sound.

$k \equiv$ specific heat ratio.

$R \equiv$ gas constant by mass unity.

$\theta \equiv$ absolute temperature.

$\lambda = \dfrac{\mu}{\rho}\sqrt{\dfrac{\pi}{2R\theta}} = \dfrac{M}{Re}\sqrt{\dfrac{k\pi}{2}} \equiv$ mean free path.

According to $\lambda$ there are three possible rarefaction grades and their corresponding motions are called:

- Continuum current (negligible rarefaction): $\dfrac{M}{Re} \ll 1$.
- Transition current (moderate rarefaction): $\dfrac{M}{Re} \approx 1$.
- Free molecular current (high rarefaction): $\dfrac{M}{Re} \gg 1$.

The motion of a body through a flow is considered, in flight dynamics, as continuum current. In the outer layer of the space that surrounds the body viscosity will be negligible. However viscosity effects will have to be considered when analysing the flow in the nearness of the boundary layer.

Depending on the compressibility of the flow, different kinds of currents can be considered:

- Incompressible subsonic current: $0 < M < 0.3$.
- Compressible subsonic current: $0.3 < M < 0.8$.
- Transonic current: $0.8 < M < 1.2$.
- Supersonic current: $1.2 < M < 5$.
- Hypersonic current: $5 < M$.

## Drag Coefficient Equation and Aerodynamic Efficiency

If the geometry of the body is fixed and in case of symmetric flight (β=0 and Q=0), pressure and friction coefficients are functions depending on:

$$C_p = C_p(\alpha, M, Re, P)$$

$$C_f = C_f(\alpha, M, Re, P)$$

where:

$\alpha \equiv$ angle of attack

$P \equiv$ considered point of the surface

Under these conditions, drag and lift coefficient are functions depending exclusively on the angle of attack of the body and Mach and Reynolds numbers. Aerodynamic efficiency, defined as the relation between lift and drag coefficients, will depend on those parameters as well.

$$\begin{cases} C_D = C_D(\alpha, M, Re) \\ C_L = C_L(\alpha, M, Re) \\ E = E(\alpha, M, Re) = \dfrac{C_L}{C_D} \end{cases}$$

It is also possible to get the dependency of the drag coefficient respect to the lift coefficient. This relation is known as the drag coefficient equation:

$C_D = C_D(C_L, M, Re) \equiv$ drag coefficient equation.

The aerodynamic efficiency has a maximum value, $E_{max}$, respect to $C_L$ where the tangent line from the coordinate origin touches the drag coefficient equation plot.

The drag coefficient, $C_D$, can be decomposed in two ways. First typical decomposition separates pressure and friction effects:

$$C_D = C_{Df} + C_{Dp} \begin{cases} C_{Df} = \dfrac{D}{qS} = -\dfrac{1}{S}\int_\Sigma C_f \mathbf{t} \bullet \mathbf{i}_\mathbf{w} \, d\sigma \\ C_{Dp} = \dfrac{D}{qS} = -\dfrac{1}{S}\int_\Sigma (-C_p) \mathbf{n} \bullet \mathbf{i}_\mathbf{w} \, d\sigma \end{cases}$$

There's a second typical decomposition taking into account the definition of the drag coefficient equation. This decomposition separates the effect of the lift coefficient in the equation, obtaining two terms $C_{Do}$ and $C_{Di}$. $C_{Do}$ is known as the parasitic drag coefficient and it is the base draft coefficient at zero lift. $C_{Di}$ is known as the induced drag coefficient and it is produced by the body lift.

$$C_D = C_{Do} + C_{Di} \begin{cases} C_{Do} = (C_D)_{C_L=0} \\ C_{Di} \end{cases}$$

## Parabolic and Generic Drag Coefficient

A good attempt for the induced drag coefficient is to assume a parabolic dependency of the lift:

$$C_{Di} = kC_L^2 \Rightarrow C_D = C_{Do} + kC_L^2$$

Aerodynamic efficiency is now calculated as:

$$E = \frac{C_L}{C_{Do} + kC_L^2} \Rightarrow \begin{cases} E_{max} = \dfrac{1}{2\sqrt{kC_{Do}}} \\ (C_L)_{Emax} = \sqrt{\dfrac{C_{Do}}{k}} \\ (C_{Di})_{Emax} = C_{Do} \end{cases}$$

If the configuration of the plane is symmetrical respect to the XY plane, minimum drag coefficient equals to the parasitic drag of the plane.

$$C_{Dmin} = (C_D)_{CL=0} = C_{Do}$$

In case the configuration is asymmetrical respect to the XY plane, however, minimum flag differs from the parasitic drag. On these cases, a new approximate parabolic drag equation can be traced leaving the minimum drag value at zero lift value.

$$C_{Dmin} = C_{DM} \neq (C_D)_{CL=0}$$

$$C_D = C_{DM} + k(C_L - C_{LM})^2$$

## Variation of Parameters with the Mach Number

The Coefficient of pressure varies with Mach number by the relation given below:

$$C_p = \frac{C_{po}}{\sqrt{|1 - M_\infty^{\ 2}|}}$$

Where:

- $C_p$ is the compressible pressure coefficient.
- $C_{po}$ is the incompressible pressure coefficient.
- $M_\infty$ is the freestream Mach number.

This relation is reasonably accurate for $0.3 < M < 0.7$ and when $M = 1$ it becomes $\infty$ which is impossible physical situation and is called Prandtl–Glauert singularity.

## Static Stability and Control

## Directional Stability

Directional or weathercock stability is concerned with the static stability of the airplane about the z axis. Just as in the case of longitudinal stability it is desirable that the aircraft should tend to return to an equilibrium condition when subjected to some form of yawing disturbance. For this the slope of the yawing moment curve must be positive. An airplane possessing this mode of stability will always point towards the relative wind, hence the name weathercock stability.

## Dynamic Stability and Control

## Longitudinal Modes

It is common practice to derive a fourth order characteristic equation to describe the longitudinal motion, and then factorize it approximately into a high frequency mode and a low frequency mode. The approach adopted here is using qualitative knowledge of aircraft behavior to simplify the equations from the outset, reaching the result by a more accessible route.

The two longitudinal motions (modes) are called the short period pitch oscillation (SPPO), and the phugoid.

## Short-period Pitch Oscillation

A short input (in control systems terminology an impulse) in pitch (generally via the elevator in a standard configuration fixed-wing aircraft) will generally lead to overshoots about the trimmed condition. The transition is characterized by a damped simple harmonic motion about the new trim. There is very little change in the trajectory over the time it takes for the oscillation to damp out.

Generally this oscillation is high frequency (hence short period) and is damped over a period of a few seconds. A real-world example would involve a pilot selecting a new climb attitude, for example 5° nose up from the original attitude. A short, sharp pull back on the control column may be used, and will generally lead to oscillations about the new trim condition. If the oscillations are poorly damped the aircraft will take a long period of time to settle at the new condition, potentially leading to Pilot-induced oscillation. If the short period mode is unstable it will generally be impossible for the pilot to safely control the aircraft for any period of time.

This damped harmonic motion is called the short period pitch oscillation, it arises from the tendency of a stable aircraft to point in the general direction of flight. It is very similar in nature to the weathercock mode of missile or rocket configurations. The motion involves mainly the pitch attitude $\theta$ (theta) and incidence $\alpha$ (alpha). The direction of the velocity vector, relative to inertial axes is $\theta-\alpha$. The velocity vector is:

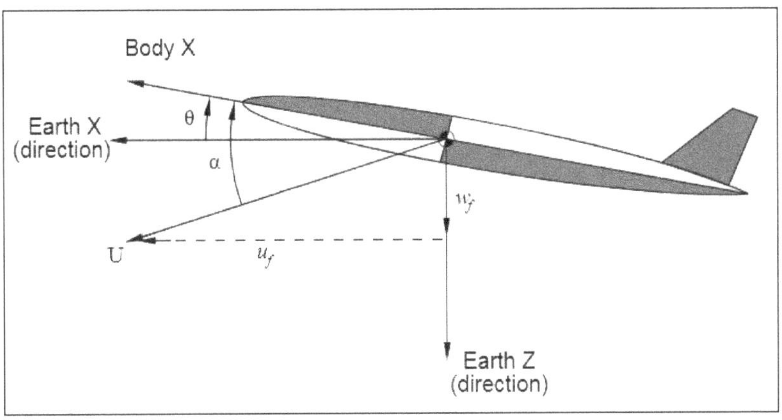


Longitudinal Equations of Motion.

$$u_f = U\cos(\theta-\alpha)$$

$$w_f = U\sin(\theta-\alpha)$$

Where $u_f, w_f$ are the inertial axes components of velocity. According to Newton's Second Law, the accelerations are proportional to the forces, so the forces in inertial axes are:

$$X_f = m\frac{du_f}{dt} = m\frac{dU}{dt}\cos(\theta-\alpha) - mU\frac{d(\theta-\alpha)}{dt}\sin(\theta-\alpha)$$

$$Z_f = m\frac{dw_f}{dt} = m\frac{dU}{dt}\sin(\theta-\alpha) + mU\frac{d(\theta-\alpha)}{dt}\cos(\theta-\alpha)$$

where $m$ is the mass. By the nature of the motion, the speed variation $m\frac{dU}{dt}$ is negligible over the period of the oscillation, so:

$$X_f = -mU\frac{d(\theta-\alpha)}{dt}\sin(\theta-\alpha)$$

$$Z_f = mU\frac{d(\theta-\alpha)}{dt}\cos(\theta-\alpha)$$

But the forces are generated by the pressure distribution on the body, and are referred to the velocity vector. But the velocity (wind) axes set is not an inertial frame so we must resolve the fixed axes forces into wind axes. Also, we are only concerned with the force along the z-axis:

$$Z = -Z_f\cos(\theta-\alpha) + X_f\sin(\theta-\alpha)$$

Or

$$Z = -mU\frac{d(\theta-\alpha)}{dt}$$

In words, the wind axes force is equal to the centripetal acceleration.

The moment equation is the time derivative of the angular momentum:

$$M = B\frac{d^2\theta}{dt^2}$$

Where M is the pitching moment, and B is the moment of inertia about the pitch axis. Let: $\frac{d\theta}{dt} = q$, the pitch rate. The equations of motion, with all forces and moments referred to wind axes are, therefore:

$$\frac{d\alpha}{dt} = q + \frac{Z}{mU}$$

$$\frac{dq}{dt} = \frac{M}{B}$$

We are only concerned with perturbations in forces and moments, due to perturbations in the states $\alpha$ and q, and their time derivatives. These are characterized by stability derivatives determined from the flight condition. The possible stability derivatives are:

- $Z_\alpha$ Lift due to incidence, this is negative because the z-axis is downwards whilst positive incidence causes an upwards force.

- $Z_q$ Lift due to pitch rate, arises from the increase in tail incidence, hence is also negative, but small compared with $Z_\alpha$.
- $M_\alpha$ Pitching moment due to incidence: The static stability term. Static stability requires this to be negative.
- $M_q$ Pitching moment due to pitch rate: The pitch damping term, this is always negative.

Since the tail is operating in the flowfield of the wing, changes in the wing incidence cause changes in the downwash, but there is a delay for the change in wing flowfield to affect the tail lift, this is represented as a moment proportional to the rate of change of incidence: $M_{\dot{\alpha}}$.

The delayed downwash effect gives the tail more lift and produces a nose down moment, so $M_{\dot{\alpha}}$ is expected to be negative.

The equations of motion, with small perturbation forces and moments become:

$$\frac{d\alpha}{dt} = \left(1 + \frac{Z_q}{mU}\right) q + \frac{Z_\alpha}{mU} \alpha$$

$$\frac{dq}{dt} = \frac{M_q}{B} q + \frac{M_\alpha}{B} \alpha + \frac{M_{\dot{\alpha}}}{B} \dot{\alpha}$$

These may be manipulated to yield as second order linear differential equation in $\alpha$:

$$\frac{d^2\alpha}{dt^2} - \left(\frac{Z_\alpha}{mU} + \frac{M_q}{B} + (1 + \frac{Z_q}{mU})\frac{M_{\dot{\alpha}}}{B}\right)\frac{d\alpha}{dt} + \left(\frac{Z_\alpha}{mU}\frac{M_q}{B} - \frac{M_\alpha}{B}(1 + \frac{Z_q}{mU})\right)\alpha = 0$$

This represents a damped simple harmonic motion.

We should expect $\frac{Z_q}{mU}$ to be small compared with unity, so the coefficient of $\alpha$ (the 'stiffness' term) will be positive, provided $M_\alpha < \frac{Z_\alpha}{mU} M_q$. This expression is dominated by $M_\alpha$, which defines the longitudinal static stability of the aircraft, it must be negative for stability. The damping term is reduced by the downwash effect, and it is difficult to design an aircraft with both rapid natural response and heavy damping. Usually, the response is underdamped but stable.

## Phugoid

If the stick is held fixed, the aircraft will not maintain straight and level flight (except in the unlikely case that it happens to be perfectly trimmed for level flight at its current altitude and thrust setting), but will start to dive, level out and climb again. It will repeat this cycle until the pilot intervenes. This long period oscillation in speed and height is called the phugoid mode. This is analyzed by assuming that the SSPO performs its proper function and maintains the angle of attack near its nominal value. The two states which are mainly affected are the climb angle $\gamma$ (gamma) and speed. The small perturbation equations of motion are:

$$mU\frac{d\gamma}{dt} = -Z$$

which means the centripetal force is equal to the perturbation in lift force.

For the speed, resolving along the trajectory:

$$m\frac{du}{dt} = X - mg\gamma$$

where g is the acceleration due to gravity at the earths surface. The acceleration along the trajectory is equal to the net x-wise force minus the component of weight. We should not expect significant aerodynamic derivatives to depend on the climb angle, so only $X_u$ and $Z_u$ need be considered. $X_u$ is the drag increment with increased speed, it is negative, likewise $Z_u$ is the lift increment due to speed increment, it is also negative because lift acts in the opposite sense to the z-axis.

The equations of motion become:

$$mU\frac{d\gamma}{dt} = -Z_u u$$

$$m\frac{du}{dt} = X_u u - mg\gamma$$

These may be expressed as a second order equation in climb angle or speed perturbation:

$$\frac{d^2u}{dt^2} - \frac{X_u}{m}\frac{du}{dt} - \frac{Z_u g}{mU}u = 0$$

Now lift is very nearly equal to weight:

$$Z = \frac{1}{2}\rho U^2 c_L S_w = W$$

where $\rho$ is the air density, $S_w$ is the wing area, W the weight and $c_L$ is the lift coefficient (assumed constant because the incidence is constant), we have, approximately:

$$Z_u = \frac{2W}{U} = \frac{2mg}{U}$$

The period of the phugoid, T, is obtained from the coefficient of u:

$$\frac{2\pi}{T} = \sqrt{\frac{2g^2}{U^2}}$$

Or

$$T = \frac{2\pi U}{\sqrt{2}g}$$

Since the lift is very much greater than the drag, the phugoid is at best lightly damped. A propeller with fixed speed would help. Heavy damping of the pitch rotation or a large rotational inertia increase the coupling between short period and phugoid modes, so that these will modify the phugoid.

## Lateral Modes

With a symmetrical rocket or missile, the directional stability in yaw is the same as the pitch stability; it resembles the short period pitch oscillation, with yaw plane equivalents to the pitch plane stability derivatives. For this reason, pitch and yaw directional stability are collectively known as the "weathercock" stability of the missile.

Aircraft lack the symmetry between pitch and yaw, so that directional stability in yaw is derived from a different set of stability derivatives. The yaw plane equivalent to the short period pitch oscillation, which describes yaw plane directional stability is called Dutch roll. Unlike pitch plane motions, the lateral modes involve both roll and yaw motion.

## Dutch Roll

It is customary to derive the equations of motion by formal manipulation in what, to the engineer, amounts to a piece of mathematical sleight of hand. The current approach follows the pitch plane analysis in formulating the equations in terms of concepts which are reasonably familiar.

Applying an impulse via the rudder pedals should induce Dutch roll, which is the oscillation in roll and yaw, with the roll motion lagging yaw by a quarter cycle, so that the wing tips follow elliptical paths with respect to the aircraft.

The yaw plane translational equation, as in the pitch plane, equates the centripetal acceleration to the side force.

$$\frac{d\beta}{dt} = \frac{Y}{mU} - r$$

where $\beta$ (beta) is the sideslip angle, Y the side force and r the yaw rate.

The moment equations are a bit trickier. The trim condition is with the aircraft at an angle of attack with respect to the airflow. The body x-axis does not align with the velocity vector, which is the reference direction for wind axes. In other words, wind axes are not principal axes (the mass is not distributed symmetrically about the yaw and roll axes). Consider the motion of an element of mass in position -z, x in the direction of the y-axis, i.e. into the plane of the paper.

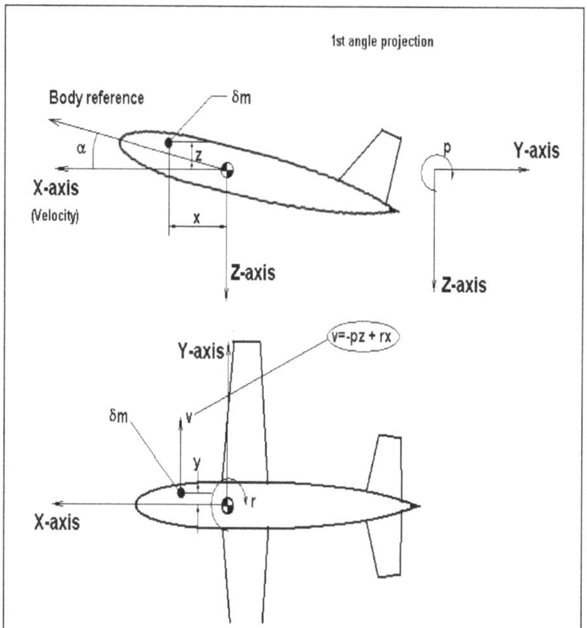


Lateral Equations Product of Inertia

If the roll rate is p, the velocity of the particle is:

$$v = -pz + xr$$

Made up of two terms, the force on this particle is first the proportional to rate of v change, the second is due to the change in direction of this component of velocity as the body moves. The latter terms gives rise to cross products of small quantities (pq, pr, qr), which are later discarded. In this analysis, they are discarded from the outset for the sake of clarity. In effect, we assume that the direction of the velocity of the particle due to the simultaneous roll and yaw rates does not change significantly throughout the motion. With this simplifying assumption, the acceleration of the particle becomes:

$$\frac{dv}{dt} = -\frac{dp}{dt}z + \frac{dr}{dt}x$$

The yawing moment is given by:

$$\delta m x \frac{dv}{dt} = -\frac{dp}{dt} xz\delta m + \frac{dr}{dt} x^2 \delta m$$

There is an additional yawing moment due to the offset of the particle in the y direction:

$$\frac{dr}{dt}y^2\delta m$$

The yawing moment is found by summing over all particles of the body:

$$N = -\frac{dp}{dt}\int xzdm + \frac{dr}{dt}\int x^2 + y^2dm = -E\frac{dp}{dt} + C\frac{dr}{dt}$$

where N is the yawing moment, E is a product of inertia, and C is the moment of inertia about the yaw axis. A similar reasoning yields the roll equation:

$$L = A\frac{dp}{dt} - E\frac{dr}{dt}$$

where L is the rolling moment and A the roll moment of inertia.

## Lateral and Longitudinal Stability Derivatives

The states are $\beta$ (sideslip), r (yaw rate) and p (roll rate), with moments N (yaw) and L (roll), and force Y (sideways). There are nine stability derivatives relevant to this motion, the following explains how they originate. However a better intuitive understanding is to be gained by simply playing with a model airplane, and considering how the forces on each component are affected by changes in sideslip and angular velocity:

- $Y_\beta$ Side force due to side slip (in absence of yaw).

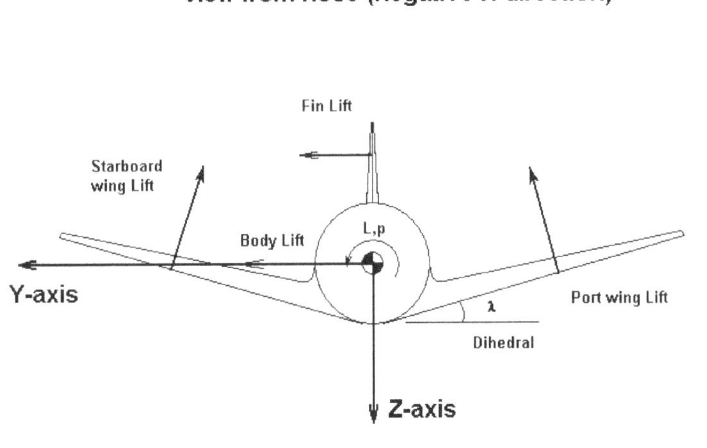


Lateral Stability - Main Sources of Stabilising Forces and Moments

Sideslip generates a sideforce from the fin and the fuselage. In addition, if the wing has dihedral, side slip at a positive roll angle increases incidence on the starboard wing and reduces it on the port side, resulting in a net force component directly opposite to the sideslip direction. Sweep back of the wings has the same effect on incidence, but since the wings are not inclined in the vertical plane, backsweep alone does not affect $Y_\beta$. However, anhedral may be used with high backsweep angles in high performance aircraft to offset the wing incidence effects of sideslip. Oddly enough this does not reverse the sign of the wing configuration's contribution to $Y_\beta$ (compared to the dihedral case).

- $Y_p$ Side force due to roll rate.

Roll rate causes incidence at the fin, which generates a corresponding side force. Also, positive roll (starboard wing down) increases the lift on the starboard wing and reduces it on the port. If the wing has dihedral, this will result in a side force momentarily opposing the resultant sideslip tendency. Anhedral wing and or stabilizer configurations can cause the sign of the side force to invert if the fin effect is swamped.

- $Y_r$ Side force due to yaw rate.

Yawing generates side forces due to incidence at the rudder, fin and fuselage.

- $N_\beta$ Yawing moment due to sideslip forces.

Sideslip in the absence of rudder input causes incidence on the fuselage and empennage, thus creating a yawing moment counteracted only by the directional stiffness which would tend to point the aircraft's nose back into the wind in horizontal flight conditions. Under sideslip conditions at a given roll angle $N_\beta$ will tend to point the nose into the sideslip direction even without rudder input, causing a downward spiraling flight.

- $N_p$ Yawing moment due to roll rate.

Roll rate generates fin lift causing a yawing moment and also differentially alters the lift on the wings, thus affecting the induced drag contribution of each wing, causing a (small) yawing moment contribution. Positive roll generally causes positive $N_p$ values unless the empennage is anhedral or fin is below the roll axis. Lateral force components resulting from dihedral or anhedral wing lift differences has little effect on $N_p$ because the wing axis is normally closely aligned with the center of gravity.

- $N_r$ Yawing moment due to yaw rate.

Yaw rate input at any roll angle generates rudder, fin and fuselage force vectors which dominate the resultant yawing moment. Yawing also increases the speed of the outboard wing whilst slowing down the inboard wing, with corresponding changes in drag causing a (small) opposing yaw moment. $N_r$ opposes the inherent directional stiffness which tends to point the aircraft's nose back into the wind and always matches the sign of the yaw rate input.

- $L_\beta$ Rolling moment due to sideslip.

A positive sideslip angle generates empennage incidence which can cause positive or negative roll moment depending on its configuration. For any non-zero sideslip angle dihedral wings causes a rolling moment which tends to return the aircraft to the horizontal, as does back swept wings. With highly swept wings the resultant rolling moment may be excessive for all stability requirements and anhedral could be used to offset the effect of wing sweep induced rolling moment.

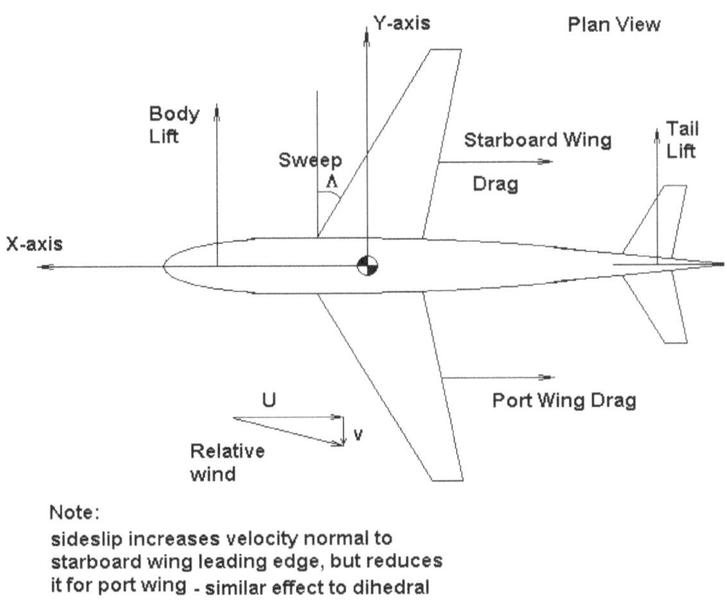


Additional Contributions to Lateral Stability

- $L_r$ Rolling moment due to yaw rate.

Yaw increases the speed of the outboard wing whilst reducing speed of the inboard one, causing a rolling moment to the inboard side. The contribution of the fin normally supports this inward rolling effect unless offset by anhedral stabilizer above the roll axis (or dihedral below the roll axis).

- $L_p$ Rolling moment due to roll rate.

Roll creates counter rotational forces on both starboard and port wings whilst also generating such forces at the empennage. These opposing rolling moment effects have to be overcome by the aileron input in order to sustain the roll rate. If the roll is stopped at a non-zero roll angle the $L_\beta$ upward rolling moment induced by the ensuing sideslip should return the aircraft to the horizontal unless exceeded in turn by the downward

$L_r$ rolling moment resulting from sideslip induced yaw rate. Longitudinal stability could be ensured or improved by minimizing the latter effect.

## Equations of Motion

Since Dutch roll is a handling mode, analogous to the short period pitch oscillation, any effect it might have on the trajectory may be ignored. The body rate $r$ is made up of the rate of change of sideslip angle and the rate of turn. Taking the latter as zero, assuming no effect on the trajectory, for the limited purpose of studying the Dutch roll:

$$\frac{d\beta}{dt} = -r$$

The yaw and roll equations, with the stability derivatives become:

$$C\frac{dr}{dt} - E\frac{dp}{dt} = N_\beta \beta - N_r \frac{d\beta}{dt} + N_p p \text{ (yaw)}$$

$$A\frac{dp}{dt} - E\frac{dr}{dt} = L_\beta \beta - L_r \frac{d\beta}{dt} + L_p p \text{ (roll)}$$

The inertial moment due to the roll acceleration is considered small compared with the aerodynamic terms, so the equations become:

$$-C\frac{d^2\beta}{dt^2} = N_\beta \beta - N_r \frac{d\beta}{dt} + N_p p$$

$$E\frac{d^2\beta}{dt^2} = L_\beta \beta - L_r \frac{d\beta}{dt} + L_p p$$

This becomes a second order equation governing either roll rate or sideslip:

$$\left(\frac{N_p}{C}\frac{E}{A} - \frac{L_p}{A}\right)\frac{d^2\beta}{dt^2} + \left(\frac{L_p}{A}\frac{N_r}{C} - \frac{N_p}{C}\frac{L_r}{A}\right)\frac{d\beta}{dt} - \left(\frac{L_p}{A}\frac{N_\beta}{C} - \frac{L_\beta}{A}\frac{N_p}{C}\right)\beta = 0$$

The equation for roll rate is identical. But the roll angle, $\phi$ (phi) is given by:

$$\frac{d\phi}{dt} = p$$

If $p$ is a damped simple harmonic motion, so is $\phi$, but the roll must be in quadrature with the roll rate, and hence also with the sideslip. The motion consists of oscillations in roll and yaw, with the roll motion lagging 90 degrees behind the yaw. The wing tips trace out elliptical paths.

Stability requires the "stiffness" and "damping" terms to be positive. These are:

$$\frac{\frac{L_p}{A}\frac{N_r}{C} - \frac{N_p}{C}\frac{L_r}{A}}{\frac{N_p}{C}\frac{E}{A} - \frac{L_p}{A}} \text{ (damping)}$$

$$\frac{\frac{L_\beta}{A}\frac{N_p}{C}-\frac{L_p}{A}\frac{N_\beta}{C}}{\frac{N_p}{C}\frac{E}{A}-\frac{L_p}{A}} \text{(stiffness)}$$

The denominator is dominated by $L_p$, the roll damping derivative, which is always negative, so the denominators of these two expressions will be positive.

Considering the "stiffness" term: $-L_pN_\beta$ will be positive because $L_p$ is always negative and $N_\beta$ is positive by design. $L_\beta$ is usually negative, whilst $N_p$ is positive. Excessive dihedral can destabilize the Dutch roll, so configurations with highly swept wings require anhedral to offset the wing sweep contribution to $L_\beta$.

The damping term is dominated by the product of the roll damping and the yaw damping derivatives, these are both negative, so their product is positive. The Dutch roll should therefore be damped.

The motion is accompanied by slight lateral motion of the center of gravity and a more "exact" analysis will introduce terms in $Y_\beta$ etc. In view of the accuracy with which stability derivatives can be calculated, this is an unnecessary pedantry, which serves to obscure the relationship between aircraft geometry and handling, which is the fundamental objective of this topic.

## Roll Subsidence

Jerking the stick sideways and returning it to center causes a net change in roll orientation.

The roll motion is characterized by an absence of natural stability, there are no stability derivatives which generate moments in response to the inertial roll angle. A roll disturbance induces a roll rate which is only canceled by pilot or autopilot intervention. This takes place with insignificant changes in sideslip or yaw rate, so the equation of motion reduces to:

$$A\frac{dp}{dt} = L_p p.$$

$L_p$ is negative, so the roll rate will decay with time. The roll rate reduces to zero, but there is no direct control over the roll angle.

## Spiral Mode

Simply holding the stick still, when starting with the wings near level, an aircraft will usually have a tendency to gradually veer off to one side of the straight flightpath. This is the (slightly unstable) spiral mode.

## Spiral Mode Trajectory

In studying the trajectory, it is the direction of the velocity vector, rather than that of the body, which is of interest. The direction of the velocity vector when projected on to the horizontal will be called the track, denoted $\mu$ (mu). The body orientation is called the heading, denoted (psi). The $\psi$ force equation of motion includes a component of weight:

$$\frac{d\mu}{dt}=\frac{Y}{mU}+\frac{g}{U}\phi$$

where $g$ is the gravitational acceleration, and $U$ is the speed.

Including the stability derivatives:

$$\frac{d\mu}{dt}=\frac{Y_\beta}{mU}\beta+\frac{Y_r}{mU}r+\frac{Y_p}{mU}p+\frac{g}{U}\phi$$

Roll rates and yaw rates are expected to be small, so the contributions of $Y_r$ and $Y_p$ will be ignored.

The sideslip and roll rate vary gradually, so their time derivatives are ignored. The yaw and roll equations reduce to:

The sideslip and roll rate vary gradually, so their time derivatives are ignored. The yaw and roll equations reduce to:

$$N_\beta\beta+N_r\frac{d\mu}{dt}+N_p p=0 \text{ (yaw)}$$

$$L_\beta\beta+L_r\frac{d\mu}{dt}+L_p p=0 \text{ (roll)}$$

Solving for $\beta$ and $p$:

$$\beta=\frac{(L_rN_p-L_pN_r)}{(L_pN_\beta-N_pL_\beta)}\frac{d\mu}{dt}$$

$$p=\frac{(L_\beta N_r-L_rN_\beta)}{(L_pN_\beta-N_pL_\beta)}\frac{d\mu}{dt}$$

Substituting for sideslip and roll rate in the force equation results in a first order equation in roll angle:

$$\frac{d\phi}{dt}=mg\frac{(L_\beta N_r-N_\beta L_r)}{mU(L_pN_\beta-N_pL_\beta)-Y_\beta(L_rN_p-L_pN_r)}\phi$$

This is an exponential growth or decay, depending on whether the coefficient of $\phi$ is positive or negative. The denominator is usually negative, which requires $L_\beta N_r > N_\beta L_r$ (both products are positive). This is in direct conflict with the Dutch roll stability requirement, and it is difficult to design an aircraft for which both the Dutch roll and spiral mode are inherently stable.

Since the spiral mode has a long time constant, the pilot can intervene to effectively stabilize it, but an aircraft with an unstable Dutch roll would be difficult to fly. It is usual to design the aircraft with a stable Dutch roll mode, but slightly unstable spiral mode.

## WEIGHT AND LIFT

The combined load of the airplane itself, the crew, the fuel, and the cargo or baggage. Weight pulls the airplane downward because of the force of gravity. It opposes lift, and acts vertically downward through the airplane's center of gravity (CG).

A fluid flowing past the surface of a body exerts a force on it. Lift is the component of this force that is perpendicular to the oncoming flow direction. It contrasts with the drag force, which is the component of the force parallel to the flow direction. Lift conventionally acts in an upward direction in order to counter the force of gravity, but it can act in any direction at right angles to the flow.

If the surrounding fluid is air, the force is called an aerodynamic force. In water or any other liquid, it is called a hydrodynamic force.

The 1902 Wright Glider shows its lift by pulling up.

Dynamic lift is distinguished from other kinds of lift in fluids. Aerostatic lift or buoyancy, in which an internal fluid is lighter than the surrounding fluid, does not require movement and is used by balloons, blimps, dirigibles, boats, and submarines. Planing lift, in which only the lower portion of the body is immersed in a liquid flow, is used by motorboats, surfboards, and water-skis.

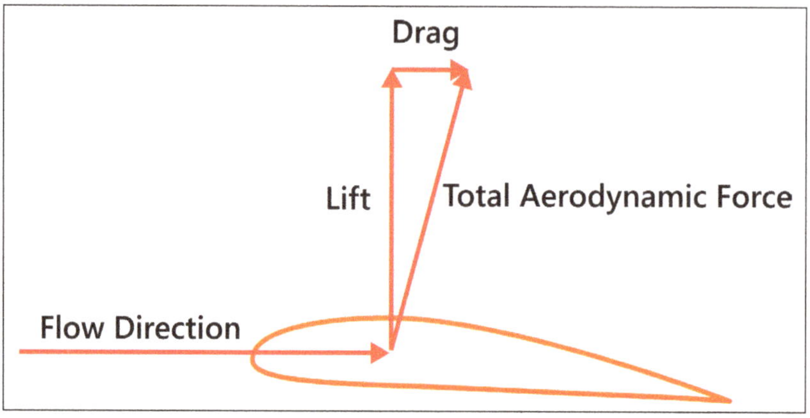


Lift is defined as the component of the aerodynamic force that is perpendicular to the flow direction, and drag is the component that is parallel to the flow direction.

A fluid flowing past the surface of a body exerts a force on it. It makes no difference whether the fluid is flowing past a stationary body or the body is moving through a stationary volume of fluid. Lift is the component of this force that is perpendicular to the oncoming flow direction. Lift is always accompanied by a drag force, which is the component of the surface force parallel to the flow direction.

Lift is mostly associated with the wings of fixed-wing aircraft, although it is more widely generated by many other streamlined bodies such as propellers, kites, helicopter rotors, racing car wings, maritime sails, and wind turbines in air, and by sailboat keels, ship's rudders, and hydrofoils in water. Lift is also exploited by flying and gliding animals, especially by birds, bats, and insects, and even in the plant world by the seeds of certain trees.

While the common meaning of the word "lift" assumes that lift opposes weight, lift can be in any direction with respect to gravity, since it is defined with respect to the direction of flow rather than to the direction of gravity. When an aircraft is cruising in straight and level flight, most of the lift opposes gravity. However, when an aircraft is climbing, descending, or banking in a turn the lift is tilted with respect to the vertical. Lift may also act as downforce in some aerobatic manoeuvres, or on the wing on a racing car. Lift may also be largely horizontal, for instance on a sailing ship.

TThe lift discussed here is mainly in relation to airfoils, although marine hydrofoils and propellers share the same physical principles and work in the same way, despite differences between air and water such as density, compressibility, and viscosity.

## Simplified Physical Explanations of Lift on an Airfoil

An airfoil is a streamlined shape that is capable of generating significantly more lift than drag. A flat plate can generate lift, but not as much as a streamlined airfoil, and with somewhat higher drag.

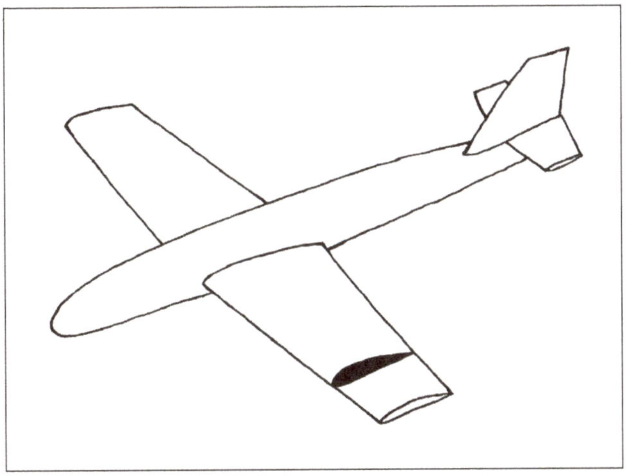

A cross-section of a wing defines an airfoil shape.

There are several ways to explain how an airfoil generates lift. Some are more complicated or more physically rigorous than others; some have been shown to be incorrect. For example, there are explanations based directly on Newton's laws of motion and explanations based on Bernoulli's principle. Either can be used to explain lift.

## Flow Deflection and Newton's Laws

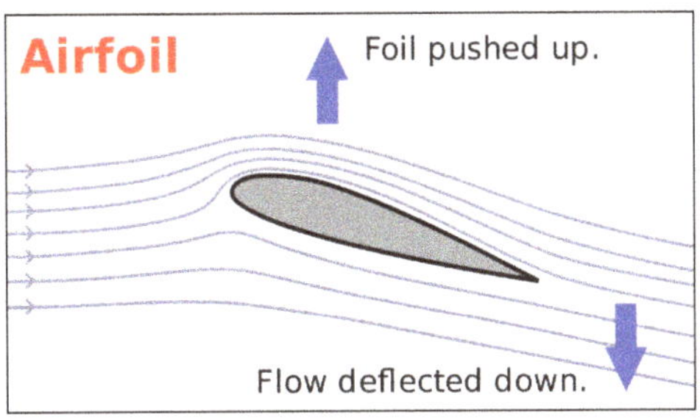


When an airfoil deflects air downwards, Newton's third law requires that the air must exert an equal upward reaction on the airfoil.

An airfoil generates lift by exerting a downward force on the air as it flows past. According to Newton's third law, the air must exert an equal and opposite (upward) force on the airfoil, which is lift.

The airflow changes direction as it passes the airfoil and follows a path that is curved downward. According to Newton's second law, this change in flow direction requires a downward force applied to the air by the airfoil. Then Newton's third law requires the air to exert an upward force on the airfoil; thus a reaction force, lift, is generated opposite to the directional change. In the case of an airplane wing, the wing exerts a downward force on the air and the air exerts an upward force on the wing.

The downward turning of the flow is not produced solely by the lower surface of the airfoil, and the air flow above the airfoil accounts for much of the downward-turning action.

## Increased Flow Speed and Bernoulli's Principle

Bernoulli's principle states that there is a direct mathematical relationship between the pressure of a fluid and the speed of that fluid, so if one knows the speed at all points within the airflow one can calculate the pressure, and vice versa. For any airfoil generating lift, there must be a pressure imbalance, i.e. lower average air pressure on the top than on the bottom. Bernoulli's principle states that this pressure difference must be accompanied by a speed difference.

## Conservation of Mass

Starting with the flow pattern observed in both theory and experiments, the increased flow speed over the upper surface can be explained in terms of streamtube pinching and conservation of mass.

For incompressible flow, the rate of volume flow (e.g. volume units per minute) must be constant within each streamtube since matter is not created or destroyed. If a streamtube becomes narrower, the flow speed must increase in the narrower region to maintain the constant flow rate, to satisfy the principle of conservation of mass.

Streamlines and streamtubes around an airfoil generating lift. Note the narrower streamtubes above and the wider streamtubes below.

The upper streamtubes constrict as they flow up and around the airfoil. Conservation of mass says that the flow speed must increase as the stream tube area decreases. Similarly, the lower streamtubes expand and their flowrate slows.

From Bernoulli's principle, the pressure on the upper surface where the flow is moving faster is lower than the pressure on the lower surface where it is moving slower. This pressure difference creates a net aerodynamic force, pointing upward.

## Limitations of the Simplified Explanations

As explained below under a more comprehensive physical explanation, producing a lift force requires maintaining pressure differences in both the vertical and horizontal directions, and thus requires both downward turning of the flow and changes in flow speed consistent with Bernoulli's principle. The simplified explanations given above are therefore incomplete because they define lift in terms of only one or the other. And depending on the details, they have other shortcomings as well.

The explanation based on flow deflection and Newton's laws is correct but is incomplete. It does not explain how the airfoil can impart downward turning to a much deeper swath of the flow than it actually touches. Further, it doesn't explain how the pressure differences in the horizontal direction are sustained. That is, it leaves out the Bernoulli part of the interaction.

Explanations based on increased flow speed and Bernoulli's principle first try to establish that there is higher flow speed over the upper surface, but they fail to explain correctly what causes the flow to speed up:

- The conservation of mass explanation that relies on narrowing of the streamtubes over the upper surface does not explain why the streamtubes change size. To see why the air flows the way it does requires more sophisticated analysis.
- Sometimes a geometrical argument is offered to demonstrate why the streamtubes change size: it is asserted that the top "obstructs" or "constricts" the air more than the bottom, hence narrower streamtubes. For conventional wings that are flat on the bottom and curved on top this makes some intuitive sense. But it does not explain how flat plates, symmetric airfoils, sailboat sails, or conventional airfoils flying upside down can generate lift, and attempts to calculate lift based on the amount of constriction do not predict experimental results.
- A common version that relies on equal-transit-time is simply wrong.

Bernoulli-only explanations imply that a speed difference arises from causes other than a pressure difference, and that the speed difference then leads to a pressure difference by Bernoulli's principle. This implied one-way causation is a misconception. The real cause-and-effect relationship between pressure and velocity is reciprocal. Finally, Bernoulli-only explanations don't explain how the pressure differences in the vertical direction are sustained. That is, they leave out the downward-turning part of the interaction.

## Alternative Explanations, Misconceptions and Controversies

Many alternative explanations for the generation of lift by an airfoil have been put forward, most intended to explain the phenomenon of lift to a general audience. Although the explanations may share features in common with the explanations above, additional assumptions and simplifications may be introduced. Some explanations introduce assumptions which proved to be wrong, such as *equal transit-time*, and some used controversial terminology, such as "Coanda effect".

## False Explanation based on Equal Transit-time

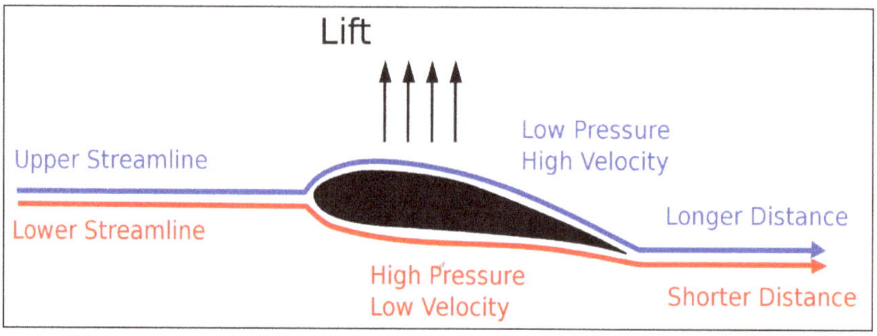


An illustration of the incorrect equal transit-time explanation of airfoil lift.

Basic or popular sources often describe the "equal transit-time" theory of lift, which incorrectly assumes that the parcels of air that divide at the leading edge of an airfoil must rejoin at the trailing edge, forcing the air traveling along the longer upper surface to go faster. Bernoulli's principle is then cited to conclude that since the air moves slower along the bottom of the wing, the air pressure must be higher, pushing the wing up.

However, there is no physical principle that requires equal transit time and experimental results show that this assumption is false. In fact, the air moving over the top of an airfoil generating lift moves *much faster* than the equal transit theory predicts. Further, the theory violates Newton's third law of motion, since it describes a force on the wing with no opposite force.

The assertion that the air must arrive simultaneously at the trailing edge is sometimes referred to as the "equal transit-time fallacy".

## Controversy Regarding the Coanda Effect

In its original sense, the Coanda effect refers to the tendency of a fluid jet to stay attached to an adjacent surface that curves away from the flow, and the resultant entrainment of ambient air into the flow. The effect is named for Henri Coanda, the Romanian aerodynamicist who exploited it in many of his patents.

More broadly, some consider the effect to include the tendency of any fluid boundary layer to adhere to a curved surface, not just the boundary layer accompanying a fluid jet. It is in this broader sense that the Coanda effect is used by some to explain why airflow remains attached to the top side of an airfoil. Jef Raskin, for example, describes a simple demonstration, using a straw to blow over the upper surface of a wing. The wing deflects upwards, thus demonstrating that the Coanda effect creates lift. This demonstration correctly demonstrates the Coanda effect as a fluid jet (the exhaust from a straw) adhering to a curved surface (the wing). However, the upper surface in this flow is a complicated, vortex-laden mixing layer, while on the lower surface the flow is quiescent. The physics of this demonstration are very different from that of the general flow over the wing. The usage in this sense is encountered in some popular references on aerodynamics. This is a controversial use of the term "Coanda effect." The more established view in the aerodynamics field is that the Coanda effect is defined in the more limited sense above, and the flow following the upper surface simply reflects an absence of boundary-layer separation; thus it is not an example of the Coanda effect.

## Basic Attributes of Lift

Lift is a result of pressure differences and depends on angle of attack, airfoil shape, air density, and airspeed.

## Pressure Differences

Pressure is the normal force per unit area exerted by the air on itself and on surfaces that it touches. The lift force is transmitted through the pressure, which acts perpendicular to the surface of the airfoil. Thus, the net force manifests itself as pressure differences. The direction of the net force implies that the average pressure on the upper surface of the airfoil is lower than the average pressure on the underside.

These pressure differences arise in conjunction with the curved airflow. When a fluid follows a curved path, there is a pressure gradient perpendicular to the flow direction with higher pressure on the outside of the curve and lower pressure on the inside. This direct relationship between curved streamlines and pressure differences, sometimes called the streamline curvature theorem, was derived from Newton's second law by Leonhard Euler in 1754:

$$\frac{\mathrm{d}p}{\mathrm{d}R} = \rho \frac{v^2}{R}$$

The left side of this equation represents the pressure difference perpendicular to the fluid flow. On the right hand side ρ is the density, v is the velocity, and R is the radius of curvature. This formula shows that higher velocities and tighter curvatures create larger pressure differentials and that for straight flow ($R \to \infty$) the pressure difference is zero.

## Angle of Attack

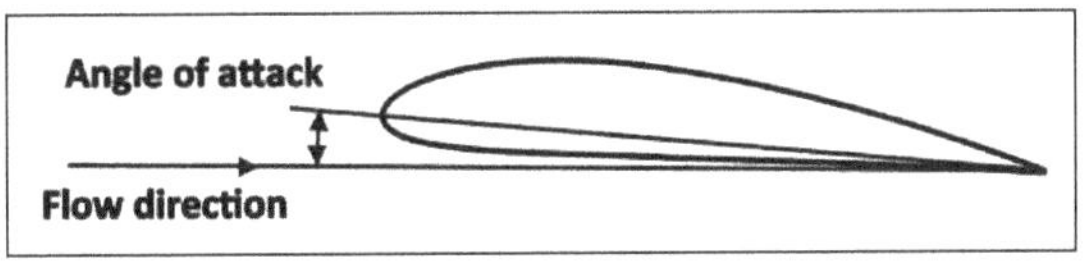


Angle of attack of an airfoil.

The angle of attack is the angle between the chord line of an airfoil and the oncoming airflow. A symmetrical airfoil will generate zero lift at zero angle of attack. But as the angle of attack increases, the air is deflected through a larger angle and the vertical component of the airstream velocity increases, resulting in more lift. For small angles a symmetrical airfoil will generate a lift force roughly proportional to the angle of attack.

As the angle of attack increases, the lift reaches a maximum at some angle; increasing the angle of attack beyond this critical angle of attack causes the upper-surface flow to separate from the wing; there is less deflection downward so the airfoil generates less lift. The airfoil is said to be stalled.

## Airfoil Shape

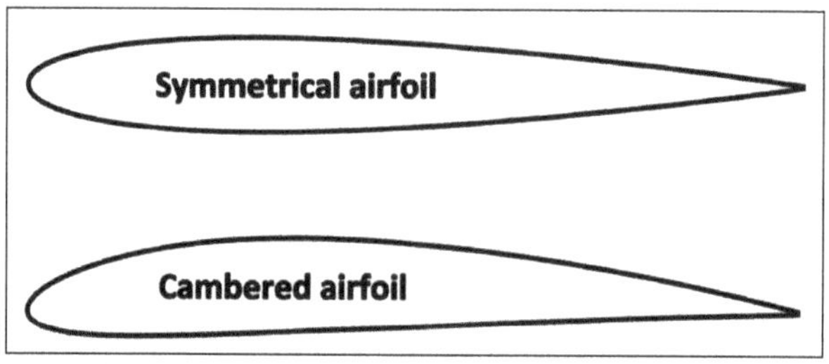


An airfoil with camber compared to a symmetrical airfoil.

The lift force depends on the shape of the airfoil, especially the amount of camber (curvature such that the upper surface is more convex than the lower surface). Increasing the camber generally increases lift.

Cambered airfoils will generate lift at zero angle of attack. When the chord line is horizontal, the trailing edge has a downward direction and since the air follows the trailing edge it is deflected downward. When a cambered airfoil is upside down, the angle of attack can be adjusted so that the lift force is upwards. This explains how a plane can fly upside down.

## Flow Conditions

The ambient flow conditions which affect lift include the fluid density, viscosity and speed of flow. Density is affected by temperature, and by the medium's acoustic velocity - i.e. by compressibility effects.

## Air Speed and Density

Lift is proportional to the density of the air and approximately proportional to the square of the flow speed. Lift also depends on the size of the wing, being generally proportional to the wing's area projected in the lift direction. In calculations it is convenient to quantify lift in terms of a lift coefficient based on these factors.

## Boundary Layer and Profile Drag

No matter how smooth the surface of an airfoil seems, any surface is rough on the scale of air molecules. Air molecules flying into the surface bounce off the rough surface in random directions relative to their original velocities. The result is that when the air is viewed as a continuous material, it is seen to be unable to slide along the surface, and the air's velocity relative to the airfoil decreases to nearly zero at the surface (i.e., the air molecules "stick" to the surface instead of sliding along it), something known as the no-slip condition. Because the air at the surface has near-zero velocity but the air away from the surface is moving, there is a thin boundary layer in which air close to the surface is subjected to a shearing motion. The air's viscosity resists the shearing, giving rise to a shear stress at the airfoil's surface called skin friction drag. Over most of the surface of most airfoils, the boundary layer is naturally turbulent, which increases skin friction drag.

Under usual flight conditions, the boundary layer remains attached to both the upper and lower surfaces all the way to the trailing edge, and its effect on the rest of the flow is modest. Compared to the predictions of inviscid flow theory, in which there is no boundary layer, the attached boundary layer reduces the lift by a modest amount and modifies the pressure distribution somewhat, which results in a viscosity-related pressure drag over and above the skin friction drag. The total of the skin friction drag and the viscosity-related pressure drag is usually called the profile drag.

## Stalling

An airfoil's maximum lift at a given airspeed is limited by boundary-layer separation. As the angle of attack is increased, a point is reached where the boundary layer can no longer remain attached to the upper surface. When the boundary layer separates, it leaves a region of recirculating flow above the upper surface, as illustrated in the flow-visualization photo at right. This is known as the *stall*, or *stalling*. At angles of attack above the stall, lift is significantly reduced, though it does not drop to zero. The maximum lift that can be achieved before stall, in terms of the lift coefficient, is generally less than 1.5 for single-element airfoils and can be more than 3.0 for airfoils with high-lift slotted flaps and leading-edge devices deployed.

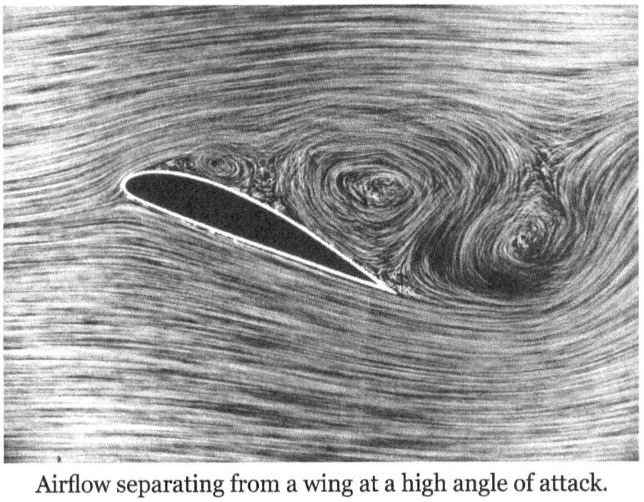

Airflow separating from a wing at a high angle of attack.

## Bluff Bodies

The flow around bluff bodies – i.e. without a streamlined shape, or stalling airfoils – may also generate lift, in addition to a strong drag force. This lift may be steady, or it may oscillate due to vortex shedding. Interaction of the object's flexibility with the vortex shedding may enhance the effects of fluctuating lift and cause vortex-induced vibrations. For instance, the flow around a circular cylinder generates a Kármán vortex street: vortices being shed in an alternating fashion from the cylinder's sides. The oscillatory nature of the flow produces a fluctuating lift force on the cylinder, even though the net (mean) force is negligible. The lift force frequency is characterised by the dimensionless Strouhal number, which depends on the Reynolds number of the flow.

For a flexible structure, this oscillatory lift force may induce vortex-induced vibrations. Under certain conditions – for instance resonance or strong spanwise correlation of the lift force – the resulting motion of the structure due to the lift fluctuations may be strongly enhanced. Such vibrations may pose problems and threaten collapse in tall man-made structures like industrial chimneys.

In the Magnus effect, a lift force is generated by a spinning cylinder in a freestream. Here the mechanical rotation acts on the boundary layer, causing it to separate at different locations on the two sides of the cylinder. The asymmetric separation changes the effective shape of the cylinder as far as the flow is concerned such that the cylinder acts like a lifting airfoil with circulation in the outer flow.

## A more Comprehensive Physical Explanation

Under "Simplified physical explanations of lift on an airfoil", there are two main popular explanations: one based on downward deflection of the flow (Newton's laws), and one based on pressure differences accompanied by changes in flow speed (Bernoulli's principle). Either of these, by itself, correctly identifies some aspects of the lifting flow but leaves other important aspects of the phenomenon unexplained.

## Lift at the Airfoil Surface

The airfoil shape and angle of attack work together so that the airfoil exerts a downward force on the air as it flows past. According to Newton's third law, the air must then exert an equal and opposite (upward) force on the airfoil, which is the lift.

The net force exerted by the air occurs as a pressure difference over the airfoil's surfaces. Pressure in a fluid is always positive in an absolute sense, so that pressure must always be thought of as pushing, and never as pulling. The pressure thus pushes inward on the airfoil everywhere on both the upper and lower surfaces. The flowing air reacts to the presence of the wing by reducing the pressure on the wing's upper surface and increasing the pressure on the lower surface. The pressure on the lower surface pushes up harder than the reduced pressure on the upper surface pushes down, and the net result is upward lift.

The pressure difference which results in lift acts directly on the airfoil surfaces; however, understanding how the pressure difference is produced requires understanding what the flow does over a wider area.

## The Wider Flow Around the Airfoil

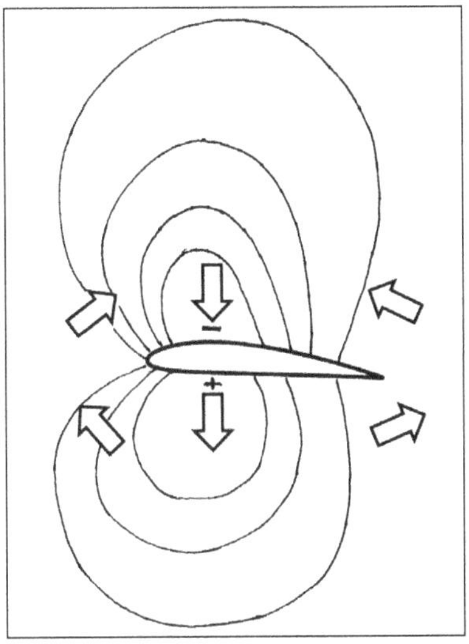


Pressure distribution with isobars around a lifting airfoil. The plus sign indicates pressure higher than ambient, and the minus sign indicates pressure lower than ambient (NOT negative pressure in the absolute sense). The block arrows indicate the directions of net forces on fluid parcels in different parts of the flowfield.

An airfoil affects the speed and direction of the flow over a wide area, producing a pattern called a velocity field. When an airfoil produces lift, the flow ahead of the airfoil is deflected upward, the flow above and below the airfoil is deflected downward, and the flow behind the airfoil is deflected upward again, leaving the air far behind the airfoil in the same state as the oncoming flow far ahead. The flow above the upper surface is sped up, while the flow below the airfoil is slowed down. Together with the upward deflection of air in front and the downward deflection of the air immediately behind, this establishes a net circulatory component of the flow. The downward deflection and the changes in flow speed are pronounced and extend over a wide area, as can be seen in the flow animation on the right. These differences in the direction and speed of the flow are greatest close to the airfoil and decrease gradually far above and below. All of these features of the velocity field also appear in theoretical models for lifting flows.

The pressure is also affected over a wide area, in a pattern of non-uniform pressure called a *pressure field*. When an airfoil produces lift, there is a diffuse region of low pressure above the airfoil, and usually a diffuse region of high pressure below, as illustrated by the isobars (curves of constant pressure) in the drawing. The pressure difference that acts on the surface is just part of this pressure field.

## Mutual Interaction of Pressure Differences and Changes in Flow Velocity

The non-uniform pressure exerts forces on the air in the direction from higher pressure to lower pressure. The direction of the force is different at different locations around the airfoil, as indicated by the block arrows in the *pressure distribution with isobars* figure. Air above the airfoil is pushed toward the center of the low-pressure region, and air below the airfoil is pushed outward from the center of the high-pressure region.

According to Newton's second law, a force causes air to accelerate in the direction of the force. Thus the vertical arrows in the pressure distribution with isobars figure indicate that air above and below the airfoil is accelerated, or turned downward, and that the non-uniform pressure is thus the cause of the downward deflection of the flow visible in the flow animation. To produce this downward turning, the airfoil must have a positive angle of attack or have its rear portion curved downward as on an airfoil with camber. Note that the downward turning of the flow over the upper surface is the result of the air being pushed downward by higher pressure above it than below it. Some explanations that refer to the "Coanda effect" suggest that viscosity plays a key role in the downward turning, but this is false.

The arrows ahead of the airfoil indicate that the flow ahead of the airfoil is deflected upward, and the arrows behind the airfoil indicate that the flow behind is deflected upward again, after being deflected downward over the airfoil. These deflections are also visible in the flow animation.

The arrows ahead of the airfoil and behind also indicate that air passing through the low-pressure region above the airfoil is sped up as it enters, and slowed back down as it leaves. Air passing through the high-pressure region below the airfoil sees the opposite - it is slowed down and then sped up. Thus the non-uniform pressure is also the cause of the changes in flow speed visible in the flow animation. The changes in flow speed are consistent with *Bernoulli's principle*, which states that in a steady flow without viscosity, lower pressure means higher speed, and higher pressure means lower speed.

Thus changes in flow direction and speed are directly caused by the non-uniform pressure. But this cause-and-effect relationship is not just one-way; it works in both directions simultaneously. The air's motion is affected by the pressure differences, but the existence of the pressure differences depends on the air's motion. The relationship is thus a mutual, or reciprocal, interaction: Air flow changes speed or direction in response to pressure differences, and the pressure differences are sustained by the air's resistance to changing speed or direction. A pressure difference can exist only if something is there for it to push against. In aerodynamic flow, the pressure difference pushes against the air's inertia, as the air is accelerated by the pressure difference. This is why the air's mass is part of the calculation, and why lift depends on air density.

Sustaining the pressure difference that exerts the lift force on the airfoil surfaces requires sustaining a pattern of non-uniform pressure in a wide area around the airfoil. This requires maintaining pressure differences in both the vertical and horizontal directions, and thus requires both downward turning of the flow and changes in flow speed according to Bernoulli's principle. The pressure differences and the changes in flow direction and speed sustain each other in a mutual interaction. The pressure differences follow naturally from Newton's second law and from the fact that flow along the surface follows the predominantly downward-sloping contours of the airfoil. And the fact that the air has mass is crucial to the interaction.

## How Simpler Explanations Fall Short

Producing a lift force requires both downward turning of the flow and changes in flow speed consistent with Bernoulli's principle. Each of the simplified explanations given above in Simplified physical explanations of lift on an airfoil falls short by trying to explain lift in terms of only one or the other, thus explaining only part of the phenomenon and leaving other parts unexplained.

## Quantifying Lift

### Pressure Integration

When the pressure distribution on the airfoil surface is known, determining the total lift requires adding up the contributions to the pressure force from local elements of the surface, each with its own local value of pressure. The total lift is thus the integral of the pressure, in the direction perpendicular to the farfield flow, over the airfoil surface.

$$L = \oint p\mathbf{n} \cdot \mathbf{k}\, \mathrm{d}S,$$

where:

- S is the projected (planform) area of the airfoil, measured normal to the mean airflow.
- n is the normal unit vector pointing into the wing.
- k is the vertical unit vector, normal to the freestream direction.

The above lift equation neglects the skin friction forces, which are small compared to the pressure forces.

By using the streamwise vector i parallel to the freestream in place of k in the integral, we obtain an expression for the pressure drag $D_p$ (which includes the pressure portion of the profile drag and, if the wing is three-dimensional, the induced drag). If we use the spanwise vector j, we obtain the side force *Y*.

$$D_p = \oint p\mathbf{n} \cdot \mathbf{i}\, dS,$$
$$Y = \oint p\mathbf{n} \cdot \mathbf{j}\, dS.$$

The validity of this integration generally requires the airfoil shape to be a closed curve that is piecewise smooth.

## Lift Coefficient

Lift depends on the size of the wing, being approximately proportional to the wing area. It is often convenient to quantify the lift of a given airfoil by its *lift coefficient* $C_L$, which defines its overall lift in terms of a unit area of the wing.

If the value of $C_L$ for a wing at a specified angle of attack is given, then the lift produced for specific flow conditions can be determined:

$$L = \tfrac{1}{2}\rho v^2 S C_L$$

where

- $L$ is the lift force.
- $\rho$ is the air density.
- $v$ is the velocity or true airspeed.
- $S$ is the planform (projected) wing area.
- $C_L$ is the lift coefficient at the desired angle of attack, Mach number, and Reynolds number.

## Mathematical Theories of Lift

Mathematical theories of lift are based on continuum fluid mechanics, assuming that air flows as a continuous fluid. Lift is generated in accordance with the fundamental principles of physics, the most relevant being the following three principles:

- Conservation of momentum, which is a consequence of Newton's laws of motion, especially Newton's second law which relates the net force on an element of air to its rate of momentum change.
- Conservation of mass, including the assumption that the airfoil's surface is impermeable for the air flowing around.
- Conservation of energy, which says that energy is neither created nor destroyed.

Because an airfoil affects the flow in a wide area around it, the conservation laws of mechanics are embodied in the form of partial differential equations combined with a set of boundary condition requirements which the flow has to satisfy at the airfoil surface and far away from the airfoil.

To predict lift requires solving the equations for a particular airfoil shape and flow condition, which generally requires calculations that are so voluminous that they are practical only on a computer, through the methods of computational fluid dynamics (CFD). Determining the net aerodynamic force from a CFD solution requires "adding up" (integrating) the forces due to pressure and shear determined by the CFD over every surface element of the airfoil as described under "pressure integration".

The Navier–Stokes equations (NS) provide the potentially most accurate theory of lift, but in practice, capturing the effects of turbulence in the boundary layer on the airfoil surface requires sacrificing some accuracy, and requires use of the Reynolds-averaged Navier–Stokes equations (RANS). Simpler but less accurate theories have also been developed.

## Navier–Stokes (NS) Equations

These equations represent conservation of mass, Newton's second law (conservation of momentum), conservation of energy, the Newtonian law for the action of viscosity, the Fourier heat conduction law, an equation of state relating density, temperature, and pressure, and formulas for the viscosity and thermal conductivity of the fluid.

In principle, the NS equations, combined with boundary conditions of no through-flow and no slip at the airfoil surface, could be used to predict lift in any situation in ordinary atmospheric flight with high accuracy. However, airflows in practical situations always involve turbulence in the boundary layer next to the airfoil surface, at least over the aft portion of the airfoil. Predicting lift by solving the NS equations in their raw form would require the calculations to resolve the details of the turbulence, down to the smallest eddy. This is not yet possible, even on the most powerful current computer. So in principle the NS equations provide a complete and very accurate theory of lift, but practical prediction of lift requires that the effects of turbulence be modeled in the RANS equations rather than computed directly.

## Reynolds-averaged Navier–Stokes (RANS) Equations

These are the NS equations with the turbulence motions averaged over time, and the effects of the turbulence on the time-averaged flow represented by turbulence modeling (an additional set of equations based on a combination of dimensional analysis and empirical information on how turbulence affects a boundary layer in a time-averaged average sense). A RANS solution consists of the time-averaged velocity vector, pressure, density, and temperature defined at a dense grid of points surrounding the airfoil.

The amount of computation required is a minuscule fraction (billionths) of what would be required to resolve all of the turbulence motions in a raw NS calculation, and with large computers available it is now practical to carry out RANS calculations for complete airplanes in three dimensions. Because turbulence models are not perfect, the accuracy of RANS calculations is imperfect, but it is adequate for practical aircraft design. Lift predicted by RANS is usually within a few percent of the actual lift.

## Inviscid-flow Equations (Euler or Potential)

The Euler equations are the NS equations without the viscosity, heat conduction, and turbulence effects. As with a RANS solution, an Euler solution consists of the velocity vector, pressure, density, and temperature defined at a dense grid of points surrounding the airfoil. While the Euler equations are simpler than the NS equations, they do not lend themselves to exact analytic solutions.

Further simplification is available through potential flow theory, which reduces the number of unknowns to be determined, and makes analytic solutions possible in some cases.

Either Euler or potential-flow calculations predict the pressure distribution on the airfoil surfaces roughly correctly for angles of attack below stall, where they might miss the total lift by as much as 10-20%. At angles of attack above stall, inviscid calculations do not predict that stall has happened, and as a result they grossly overestimate the lift.

In potential-flow theory, the flow is assumed to be irrotational, i.e. that small fluid parcels have no net rate of rotation. Mathematically, this is expressed by the statement that the curl of the velocity vector field is everywhere equal to zero. Irrotational flows have the convenient property that the velocity can be expressed as the gradient of a scalar function called a potential. A flow represented in this way is called potential flow.

In potential-flow theory, the flow is assumed to be incompressible. Incompressible potential-flow theory has the advantage that the equation (Laplace's equation) to be solved for the potential is linear, which allows solutions to be constructed by superposition of other known solutions. The incompressible-potential-flow equation can also be solved by conformal mapping, a method based on the theory of functions of a complex variable. In the early 20th century, before computers were available, conformal mapping was used to generate solutions to the incompressible potential-flow equation for a class of idealized airfoil shapes, providing some of the first practical theoretical predictions of the pressure distribution on a lifting airfoil.

A solution of the potential equation directly determines only the velocity field. The pressure field is deduced from the velocity field through Bernoulli's equation.

Applying potential-flow theory to a lifting flow requires special treatment and an additional assumption. The problem arises because lift on an airfoil in inviscid flow requires circulation in the flow around the airfoil, but a single potential function that is continuous throughout the domain around the airfoil cannot represent a flow with nonzero circulation. The solution to this problem is to introduce a branch cut, a curve or line from some point on the airfoil surface out to infinite distance, and to allow a jump in the value of the potential across the cut. The jump in the potential imposes circulation in the flow equal to the potential jump and thus allows nonzero circulation to be represented. However, the potential jump is a free parameter that is not determined by the potential equation or the other boundary conditions, and the solution is thus indeterminate. A potential-flow solution exists for any value of the circulation and any value of the lift. One way to resolve this indeterminacy is to impose the Kutta condition, which is that, of all the possible solutions, the physically reasonable solution is the one in which the flow leaves the trailing edge smoothly. The streamline sketches illustrate one flow pattern with zero lift, in which the flow goes around the trailing edge and leaves the upper surface ahead of the trailing edge, and another flow pattern with positive lift, in which the flow leaves smoothly at the trailing edge in accordance with the Kutta condition.

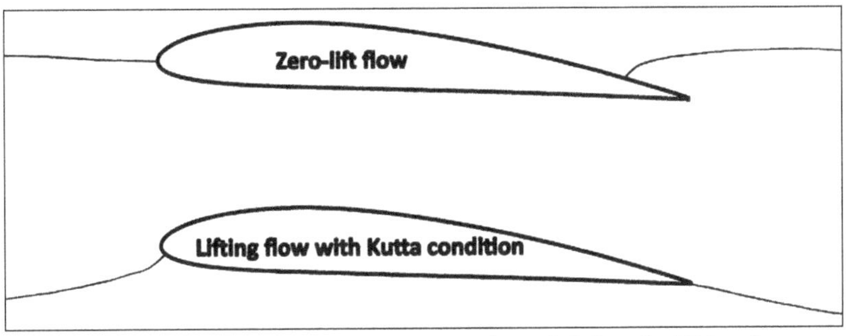


Comparison of a non-lifting flow pattern around an airfoil and a lifting flow pattern consistent with the Kutta condition, in which the flow leaves the trailing edge smoothly.

## Linearized Potential Flow

This is potential-flow theory with the further assumptions that the airfoil is very thin and the angle of attack is small. The linearized theory predicts the general character of the airfoil pressure distribution and how it is influenced by airfoil shape and angle of attack, but is not accurate enough for design work. For a 2D airfoil, such calculations can be done in a fraction of a second in a spreadsheet on a PC.

## Circulation and the Kutta–Joukowski Theorem

When an airfoil generates lift, several components of the overall velocity field contribute to a net circulation of air around it: the upward flow ahead of the airfoil, the accelerated flow above, the decelerated flow below, and the downward flow behind.

The circulation can be understood as the total amount of "spinning" (or vorticity) of an inviscid fluid around the airfoil.

The Kutta–Joukowski theorem relates the lift per unit width of span of a two-dimensional airfoil to this circulation component of the flow. It is a key element in an explanation of lift that follows the development of the flow around an airfoil as the airfoil starts its motion from rest and a starting vortex is formed and left behind, leading to the formation of circulation around the airfoil. Lift is then inferred from the Kutta-Joukowski theorem. This explanation is largely mathematical, and its general progression is based on logical inference, not physical cause-and-effect.

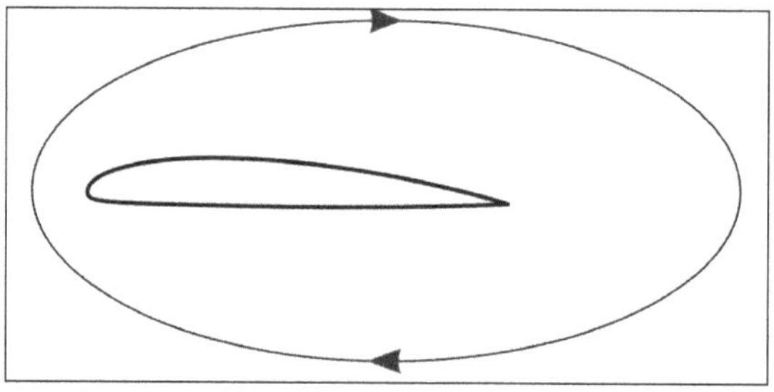

Circulation component of the flow around an airfoil.

The Kutta–Joukowski model does not predict how much circulation or lift a two-dimensional airfoil will produce. Calculating the lift per unit span using Kutta–Joukowski requires a known value for the circulation. In particular, if the Kutta condition is met, in which the rear stagnation point moves to the airfoil trailing edge and attaches there for the duration of flight, the lift can be calculated theorically through the conformal mapping method.

The lift generated by a conventional airfoil is dictated by both its design and the flight conditions, such as forward velocity, angle of attack and air density. Lift can be increased by artificially increasing the circulation, for example by boundary-layer blowing or the use of blown flaps. In the Flettner rotor the entire airfoil is circular and spins about a spanwise axis to create the circulation.

## Three-dimensional Flow

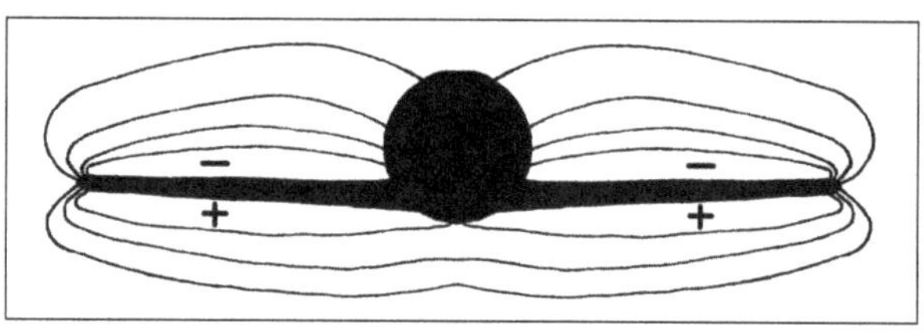


Cross-section of an airplane wing-body combination showing the isobars of the three-dimensional lifting flow.

Cross-section of an airplane wing-body combination showing velocity vectors of the three-dimensional lifting flow.

The flow around a three-dimensional wing involves significant additional issues, especially relating to the wing tips. For a wing of low aspect ratio, such as a typical delta wing, two-dimensional theories may provide a poor model and three-dimensional flow effects can dominate. Even for wings of high aspect ratio, the three-dimensional effects associated with finite span can affect the whole span, not just close to the tips.

## Wing Tips and Spanwise Distribution

The vertical pressure gradient at the wing tips causes air to flow sideways, out from under the wing then up and back over the upper surface. This reduces the pressure gradient at the wing tip, therefore also reducing lift. The lift tends to decrease in the spanwise direction from root to tip, and the pressure distributions around the airfoil sections change accordingly in the spanwise direction. Pressure distributions in planes perpendicular to the flight direction tend to look like the illustration at right. This spanwise-varying pressure distribution is sustained by a mutual interaction with the velocity field. Flow below the wing is accelerated outboard, flow outboard of the tips is accelerated upward, and flow above the wing is accelerated inboard, which results in the flow pattern illustrated at right.

There is more downward turning of the flow than there would be in a two-dimensional flow with the same airfoil shape and sectional lift, and a higher sectional angle of attack is required to achieve the same lift compared to a two-dimensional flow. The wing is effectively flying in a downdraft of its own making, as if the freestream flow were tilted downward, with the result that the total aerodynamic force vector is tilted backward slightly compared to what it would be in two dimensions. The additional backward component of the force vector is called lift-induced drag.

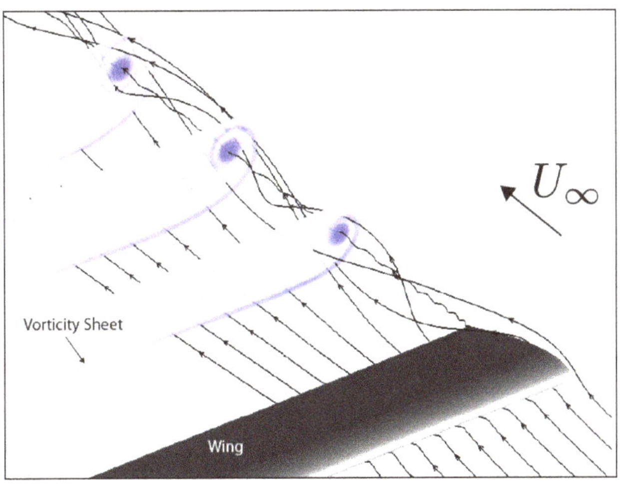


Euler computation of a tip vortex rolling up from the trailed vorticity sheet.

The difference in the spanwise component of velocity above and below the wing (between being in the inboard direction above and in the outboard direction below) persists at the trailing edge and into the wake downstream. After the flow leaves the trailing edge, this difference in velocity takes place across a relatively thin shear layer called a vortex sheet.

## Horseshoe Vortex System

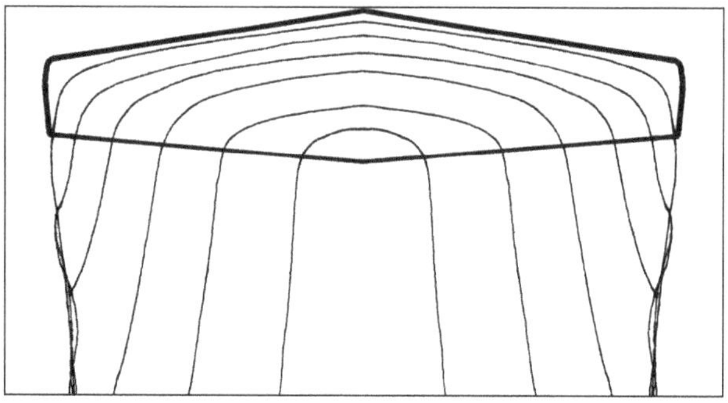

Planview of a wing showing the horseshoe vortex system.

The wingtip flow leaving the wing creates a tip vortex. As the main vortex sheet passes downstream from the trailing edge, it rolls up at its outer edges, merging with the tip vortices. The combination of the wingtip vortices and the vortex sheets feeding them is called the vortex wake.

In addition to the vorticity in the trailing vortex wake there is vorticity in the wing's boundary layer, called 'bound vorticity', which connects the trailing sheets from the two sides of the wing into a vortex system in the general form of a horseshoe. The horseshoe form of the vortex system was recognized by the British aeronautical pioneer Lanchester in 1907.

Given the distribution of bound vorticity and the vorticity in the wake, the Biot–Savart law (a vector-calculus relation) can be used to calculate the velocity perturbation anywhere in the field, caused by the lift on the wing. Approximate theories for the lift distribution and lift-induced drag of three-dimensional wings are based on such analysis applied to the wing's horseshoe vortex system. In these theories, the bound vorticity is usually idealized and assumed to reside at the camber surface inside the wing.

Because the velocity is deduced from the vorticity in such theories, some authors describe the situation to imply that the vorticity is the cause of the velocity perturbations, using terms such as "the velocity induced by the vortex", for example But attributing mechanical cause-and-effect between the vorticity and the velocity in this way is not consistent with the physics. The velocity perturbations in the flow around a wing are in fact produced by the pressure field.

## Manifestations of Lift in the Farfield

## Integrated Force/Momentum Balance in Lifting Flows

Control volumes of different shapes that have been used in analyzing the momentum balance in the 2D flow around a lifting airfoil. The airfoil is assumed to exert a downward force -L' per unit span on the air, and the proportions in which that force is manifested as momentum fluxes and pressure differences at the outer boundary are indicated for each different shape of control volume.

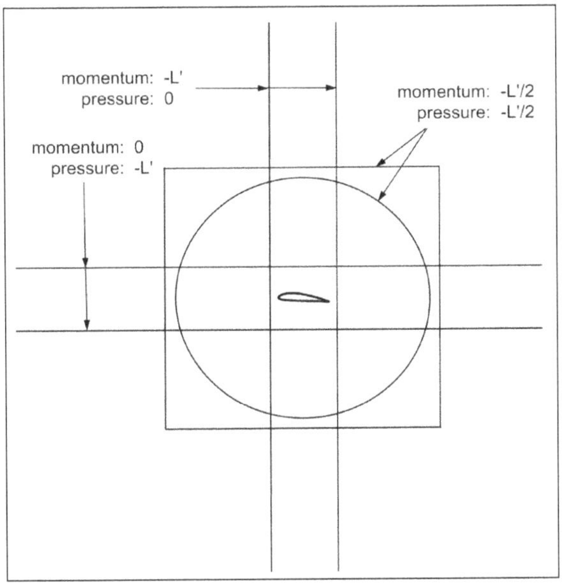


The flow around a lifting airfoil must satisfy Newton's second law regarding conservation of momentum, both locally at every point in the flow field, and in an integrated sense over any extended region of the flow. For an extended region, Newton's second law takes the form of the *momentum theorem for a control volume*, where a control volume can be any region of the flow chosen for analysis. The momentum theorem states that the integrated force exerted at the boundaries of the control volume (a surface integral), is equal to the integrated time rate of change (material derivative) of the momentum of fluid parcels passing through the interior of the control volume. For a steady flow, this can be expressed in the form of the net surface integral of the flux of momentum through the boundary.

The lifting flow around a 2D airfoil is usually analyzed in a control volume that completely surrounds the airfoil, so that the inner boundary of the control volume is the airfoil surface, where the downward force per unit span $-L'$ is exerted on the fluid by the airfoil. The outer boundary is usually either a large circle or a large rectangle. At this outer boundary distant from the airfoil, the velocity and pressure are well represented by the velocity and pressure associated with a uniform flow plus a vortex, and viscous stress is negligible, so that the only force that must be integrated over the outer boundary is the pressure. The free-stream velocity is usually assumed to be horizontal, with lift vertically upward, so that the vertical momentum is the component of interest.

For the free-air case (no ground plane), the force $-L'$ exerted by the airfoil on the fluid is manifested partly as momentum fluxes and partly as pressure differences at the outer boundary, in proportions that depend on the shape of the outer boundary, as shown in the diagram. For a flat horizontal rectangle that is much longer than it is tall, the fluxes of vertical momentum through the front and back are negligible, and the lift is accounted for entirely by the integrated pressure differences on the top and bottom. For a square or circle, the momentum fluxes and pressure differences account for half the lift each. For a vertical rectangle that is much taller than it is wide, the unbalanced pressure forces on the top and bottom are negligible, and lift is accounted for entirely by momentum fluxes, with a flux of upward momentum that enters the control volume through the front accounting for half the lift, and a flux of downward momentum that exits the control volume through the back accounting for the other half.

The results of all of the control-volume analyses described above are consistent with the Kutta–Joukowski theorem described above. Both the tall rectangle and circle control volumes have been used in derivations of the theorem.

## Lift Reacted by Overpressure on the Ground under an Airplane

An airfoil produces a pressure field in the surrounding air, as explained under "The wider flow around the airfoil" above. The pressure differences associated with this field die off gradually, becoming very small at large distances, but never disappearing altogether. Below the airplane, the pressure field persists as a positive pressure disturbance that reaches the ground, forming a pattern of slightly-higher-than-ambient pressure on the ground, as shown on the right. Although the pressure differences are very small far below the airplane, they are spread over a wide area and add up to a substantial force. For steady, level flight, the integrated force due to the pressure differences is equal to the total aerodynamic lift of the airplane and to the airplane's weight. According to Newton's third law, this pressure force exerted on the ground by the air is matched by an equal-and-opposite upward force exerted on the air by the ground, which offsets all of the downward force exerted on the air by the airplane. The net force due to the lift, acting on the atmosphere as a whole, is therefore zero, and thus there is no integrated accumulation of vertical momentum in the atmosphere, as was noted by Lanchester early in the development of modern aerodynamics.

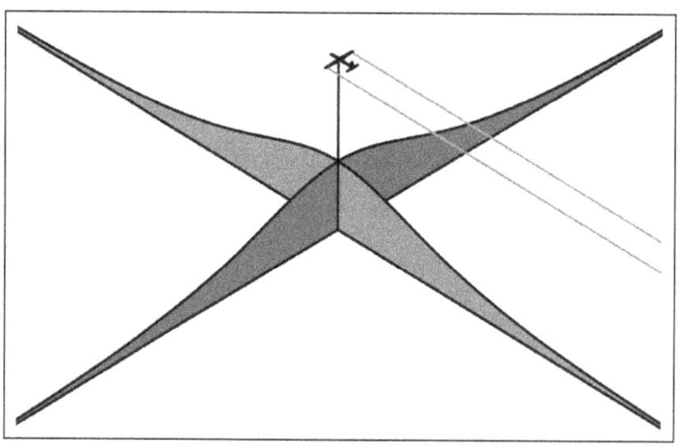

Illustration of the distribution of higher-than-ambient pressure on the ground under an airplane in flight.

# THRUST

A Pratt & Whitney F100 jet engine being tested. This engine produces a jet of gas to generate thrust. Its purpose is to propel a jet airplane.

Thrust is a reaction force described quantitatively by Newton's third law. When a system expels or accelerates mass in one direction, the accelerated mass will cause a force of equal magnitude but opposite direction on that system. The force applied on a surface in a direction perpendicular or normal to the surface is also called thrust. Force, and thus thrust, is measured using the International System of Units (SI) in newtons (symbol: N), and represents the amount needed to accelerate 1 kilogram of mass at the rate of 1 meter per second per second. In mechanical engineering, force orthogonal to the main load (such as in parallel helical gears) is referred to as thrust.

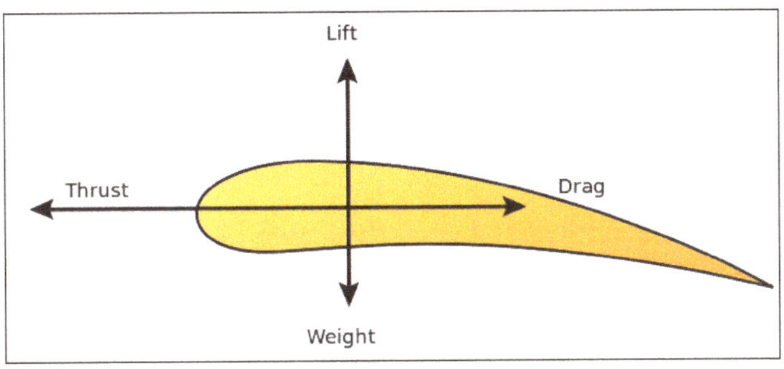


Forces on an aerofoil cross section.

A fixed-wing aircraft generates forward thrust when air is pushed in the direction opposite to flight. This can be done in several ways including by the spinning blades of a propeller, or a rotating fan pushing air out from the back of a jet engine, or by ejecting hot gases from a rocket engine. The forward thrust is proportional to the mass of the airstream multiplied by the difference in velocity of the airstream. Reverse thrust can be generated to aid braking after landing by reversing the pitch of variable-pitch propeller blades, or using a thrust reverser on a jet engine. Rotary wing aircraft and thrust vectoring V/STOL aircraft use engine thrust to support the weight of the aircraft, and vector sum of this thrust fore and aft to control forward speed.

A motorboat generates thrust (or reverse thrust) when the propellers are turned to accelerate water backwards (or forwards). The resulting thrust pushes the boat in the opposite direction to the sum of the momentum change in the water flowing through the propeller.

A rocket is propelled forward by a thrust force equal in magnitude, but opposite in direction, to the time-rate of momentum change of the exhaust gas accelerated from the combustion chamber through the rocket engine nozzle. This is the exhaust velocity with respect to the rocket, times the time-rate at which the mass is expelled, or in mathematical terms:

$$\mathbf{T} = \mathbf{v}\frac{dm}{dt}$$

Where T is the thrust generated (force), $\frac{dm}{dt}$ is the rate of change of mass with respect to time (mass flow rate of exhaust), and v is the speed of the exhaust gases measured relative to the rocket.

For vertical launch of a rocket the initial thrust at liftoff must be more than the weight.

Each of the three Space Shuttle Main Engines could produce a thrust of 1.8 meganewton, and each of the Space Shuttle's two Solid Rocket Boosters 14.7 MN (3,300,000 lbf), together 29.4 MN.

By contrast, the simplified Aid For EVA Rescue (SAFER) has 24 thrusters of 3.56 N (0.80 lbf) each.

## Concepts

### Thrust to Power

The power needed to generate thrust and the force of the thrust can be related in a non-linear way. In general, $\mathbf{P}^2 \propto \mathbf{T}^3$. The proportionality constant varies, and can be solved for a uniform flow:

$$\frac{dm}{dt} = \rho A v$$

$$\mathbf{T} = \frac{dm}{dt}v, \mathbf{P} = \frac{1}{2}\frac{dm}{dt}v^2$$

$$\mathbf{T} = \rho A v^2, \mathbf{P} = \frac{1}{2}\rho A v^3$$

$$P^2 = \frac{\mathbf{T}^3}{4\rho A}$$

Note that these calculations are only valid for when the incoming air is accelerated from a standstill – for example when hovering.

The inverse of the proportionality constant, the "efficiency" of an otherwise-perfect thruster, is proportional to the area of the cross section of the propelled volume of fluid ($A$) and the density of the fluid ($\rho$). This helps to explain why moving through water is easier and why aircraft have much larger propellers than watercraft.

## Thrust to Propulsive Power

A very common question is how to compare the thrust rating of a jet engine with the power rating of a piston engine. Such comparison is difficult, as these quantities are not equivalent. A piston engine does not move the aircraft by itself (the propeller does that), so piston engines are usually rated by how much power they deliver to the propeller. Except for changes in temperature and air pressure, this quantity depends basically on the throttle setting.

A jet engine has no propeller, so the propulsive power of a jet engine is determined from its thrust as follows. Power is the force (F) it takes to move something over some distance (d) divided by the time (t) it takes to move that distance:

$$\mathbf{P} = \mathbf{F}\frac{d}{t}$$

In case of a rocket or a jet aircraft, the force is exactly the thrust (T) produced by the engine. If the rocket or aircraft is moving at about a constant speed, then distance divided by time is just speed, so power is thrust times speed:

$$\mathbf{P} = \mathbf{T}v$$

This formula looks very surprising, but it is correct: the *propulsive power* (or *power available*) of a jet engine increases with its speed. If the speed is zero, then the propulsive power is zero. If a jet aircraft is at full throttle but attached to a static test stand, then the jet engine produces no propulsive power, however thrust is still produced. The combination piston engine–propeller also has a propulsive power with exactly the same formula, and it will also be zero at zero speed – but that is for the engine–propeller set. The engine alone will continue to produce its rated power at a constant rate, whether the aircraft is moving or not.

Now, imagine the strong chain is broken, and the jet and the piston aircraft start to move. At low speeds:

- The piston engine will have constant 100% power, and the propeller's thrust will vary with speed.
- The jet engine will have constant 100% thrust, and the engine's power will vary with speed.

### Excess Thrust

If a powered aircraft is generating thrust T and experiencing drag D, the difference between the two, T – D, is termed the excess thrust. The instantaneous performance of the aircraft is mostly dependent on the excess thrust.

Excess thrust is a vector and is determined as the vector difference between the thrust vector and the drag vector.

### Thrust Axis

The thrust axis for an airplane is the line of action of the total thrust at any instant. It depends on the location, number, and characteristics of the jet engines or propellers. It usually differs from the drag axis. If so, the distance between the thrust axis and the drag axis will cause a moment that must be resisted by a change in the aerodynamic force on the horizontal stabiliser. Notably, the Boeing 737 MAX, with larger, lower-slung engines than previous 737 models, had a greater distance between the thrust axis and the drag axis, causing the nose to rise up in some flight regimes, necessitating a pitch-control system, MCAS.

## DRAG

Any physical body being propelled through the air has drag associated with it. In aerodynamics, drag is defined as the force that opposes forward motion through the atmosphere and is parallel to the direction of the free-stream velocity of the airflow. Drag must be overcome by thrust in order to achieve forward motion.

Drag is generated by nine conditions associated with the motion of air particles over the aircraft. There are several types of drag: form, pressure, skin friction, parasite, induced, and wave.

### Induced Drag

Induced drag is the drag created by the vortices at the tip of an aircraft's wing. Induced drag is the drag due to lift. The high pressure underneath the wing causes the airflow at the tips of the wings to curl around from bottom to top in a circular motion. This results in a trailing vortex. Induced drag increases in direct proportion to increases in the angle of attack. The circular motion creates a change in the angle of attack near the wing tip which causes an increase in drag. The greater the angle of attack up to the critical angle (where a stall takes place), the greater the amount of lift developed and the greater the induced drag.

All of these types of drag must be accounted for when determining drag for subsonic flight. The total drag is the sum of parasite and induced drag.

Total Drag = Parasite Drag + Induced Drag

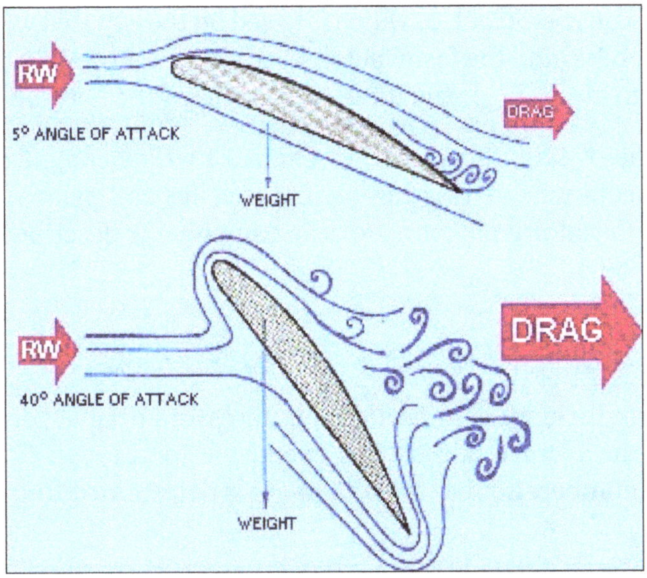


Induced drag is a by-product of lift.

But the net (or total) drag of an aircraft is not simply the sum of the drag of its components. When the components are combined into a complete aircraft, one component can affect the air flowing around and over the airplane, and hence, the drag of one component can affect the drag associated with another component. These effects are called *interference effects*, and the change in the sum of the component drags is called *interference drag*. Thus,

$$(\text{Drag})_{1+2} = (\text{Drag})_1 + (\text{Drag})_2 + (\text{Drag})_{\text{interference}}$$

Generally, interference drag will add to the component drags but in a few cases, for example, adding tip tanks to a wing, total drag will be less than the sum of the two component drags because of the reduction of induced drag.

## Parasite Drag

The parasite drag of a typical airplane in the cruise configuration consists primarily of the skin friction, roughness, and pressure drag of the major components. There is usually some additional parasite drag due to such things as fuselage upsweep, control surface gaps, base areas, and other extraneous items. Since most of the elements that make up the total parasite drag are dependent on Reynolds number and since some are dependent on Mach number, it is necessary to specify the conditions under which the parasite drag is to be evaluated. In the method of these notes, the conditions selected are the Mach number and the Reynolds number corresponding to the flight condition of interest.

The basic parasite drag area for airfoil and body shapes can be computed from the following expression:

$$f = k\, c_f\, S_{wet}$$

where the skin friction coefficient, $c_f$, which is based on the exposed wetted area includes the effects of roughness, and the form factor, k, accounts for the effects of both supervelocities and pressure drag. $S_{wet}$ is the total wetted area of the body or surface.

Computation of the overall parasite drag requires that we compute the drag area of each of the major components (fuselage, wing, nacelles and pylons, and tail surfaces) and then evaluate the additional parasite drag components described above.

We thus write:

$$C_{Dp} = S\, k_i\, c_{fi}\, S_{weti}/S_{ref} + C_{Dupsweep} + C_{Dgap} + C_{Dnac\_base} + C_{Dmisc}$$

where the first term includes skin friction, and pressure drag at zero lift of the major components. $c_{fi}$ is the average skin friction coefficient for a rough plate with transition at flight Reynolds number. Equivalent roughness is determined from flight test data.

## Form Drag

Form drag and pressure drag are virtually the same type of drag. Form or pressure drag is caused by the air that is flowing over the aircraft or airfoil. The separation of air creates turbulence and results in pockets of low and high pressure that leave a wake behind the airplane or airfoil (thus the name pressure drag). This opposes forward motion and is a component of the total drag. Since this drag is due to the shape, or form of the aircraft, it is also called form drag. Streamlining the aircraft will reduce form drag, and parts of an aircraft that do not lend themselves to streamlining are enclosed in covers called fairings, or a cowling for an engine, that have a streamlined shape. Airplane components that produce form drag include (1) the wing and wing flaps, (2) the fuselage, (3) tail surfaces, (4) nacelles, (5) landing gear, (6) wing tanks and external stores, and (7) engines.

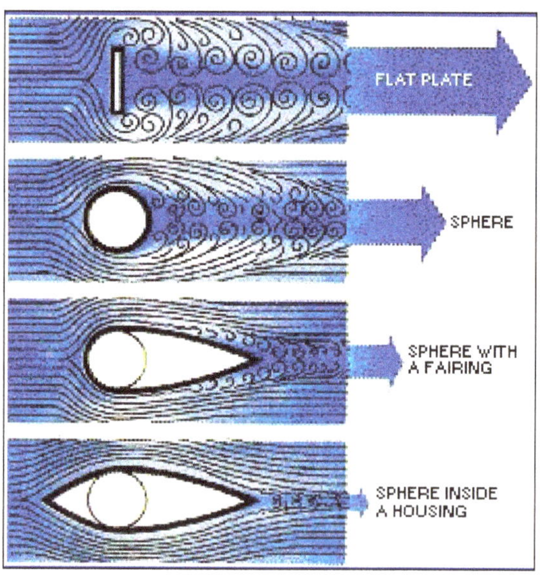

Skin friction drag is caused by the actual contact of the air particles against the surface of the aircraft. This is the same as the friction between any two objects or substances. Because skin friction drag is an interaction between a solid (the airplane surface) and a gas (the air), the magnitude of skin friction drag depends on the properties of both the solid and the gas. For the solid airplane, skin fiction drag can be reduced, and airspeed can be increased somewhat, by keeping an aircraft's surface highly polished and clean. For the gas, the magnitude of the drag depends on the viscosity of the air. Along the solid surface of the airplane, a boundary layer of low energy flow is generated. The magnitude of the skin friction depends on the state of this flow.

## Skin Friction

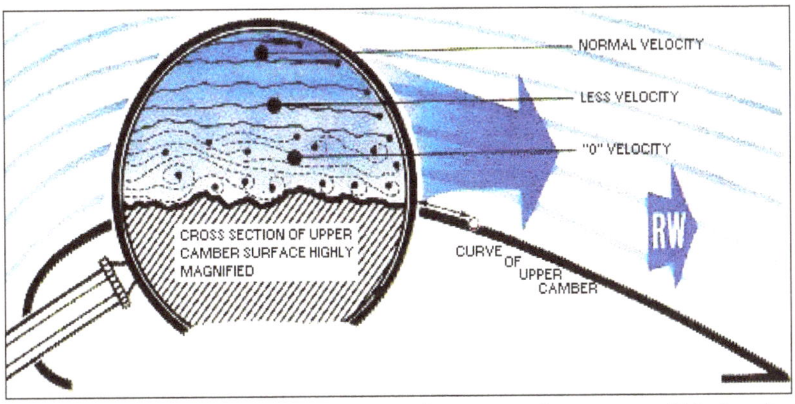


The leading edge of a wing will always produce a certain amount of friction drag.

An important aerodynamic force during low-speed subsonic flight is the shear force (the sideways force or internal friction) caused by viscous airflow over the surfaces of the vehicle. This shear force is referred to as the skin-friction force or skin-friction drag and depends strongly on the Reynolds number, surface roughness, and pressure gradients. In addition to the pressure forces that act everywhere perpendicular to (normal to) a body in moving air, viscous forces are also present. It is these viscous forces that modify the lift that would exist under ideal conditions (air is inviscid and incompressible) and help create the real drag.

If the airflow were ideal, that is, inviscid, the air would simply slip over the surface of a smooth plate with velocity V¥. At all points along the surface of the plate, the velocity distribution (that is, the variation in velocity as one moves perpendicularly away from the surface) would be a uniform constant value of $V_\infty$. No drag would result if the airflow were frictionless (inviscid).

Under real conditions, however, a very thin film of air molecules adheres to the surface. This is the very important no-slip condition. It states that at the surface of a body, the airflow velocity is zero. As one moves away from the body, the velocity of the air gradually increases until, at some point, the velocity becomes a constant value; in the case of a flat plate this value is V¥. The layer of air where the velocity is changing from zero to a constant value is called the boundary layer. Within the boundary layer, there are relative velocities between the layers and an internal friction is present. This internal friction extends to the surface of the body. The cumulative effect of all these friction forces is to produce drag on the plate. This drag is referred to as skin-friction drag.

Initially, near the leading edge of a flat, smooth plate, one has a laminar flow (the flow is layered) and the boundary layer also is steady and layered—hence, a laminar boundary layer. As one moves farther downstream, viscosity continues to act, and the laminar boundary layer thickens as more and more air is slowed down by internal friction. Eventually, a point is reached on the plate where the laminar boundary layer undergoes transition and becomes a turbulent boundary layer. As is usual for turbulent flow, there is random motion in the boundary layer as well as the downstream-directed motion. There is no slip at the surface of the plate. Another important difference from the laminar boundary layer is the fact that the velocity builds up more quickly as one moves away from the wall, although the total boundary-layer thickness is greater. The turbulent boundary layer farther away from the wall reenergizes the slower moving air nearer the wall. This condition can be seen by comparing the profile of the laminar boundary layer with the profile of the turbulent boundary layer.

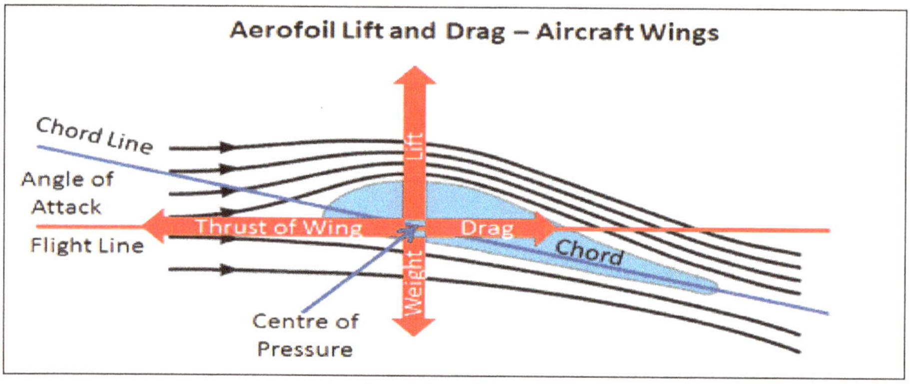


Aerodynamics of flight drag.

The Reynolds number has an important effect on the boundary layer. As the Reynolds number increases (caused by increasing the airflow speed and/or decreasing the viscosity), the boundary layer thickens more slowly. However, even though the Reynolds number becomes large, the velocity at the surface of the body must be zero. Thus, the boundary layer never disappears.

It is interesting to note that a typical thickness of the boundary layer on an aircraft wing is generally less than a centimetre (2.5 inches). Yet, the velocity must vary from zero at the surface of the wing to hundreds of meters per second at the outer edge of the boundary layer. It is evident that tremendous shearing forces (internal friction) must be acting in this region. This gives rise to the skin-friction drag.

Applied to an airfoil in a real airflow, the same free-stream velocity V¥ and free-stream static pressure p¥ apply. The field of air ahead of the airfoil is only slightly modified and for all practical purposes, the velocities and static pressures are the same as for the ideal fluid case. Again a stagnation point (a point with no motion) occurs at the leading edge of the airfoil and the pressure reaches its maximum value of pt at this point (total or stagnation pressure). From this point on along the airfoil, the picture changes.

On a flat plate (for example), a boundary layer begins to form because of viscosity. This boundary layer is very thin and outside of it, the flow acts very much like that of an ideal fluid. Also, the static pressure acting on the surface of the airfoil is determined by the static pressure outside the boundary layer. This pressure is transmitted through the boundary layer to the surface and thus acts as if the boundary layer were not present at all. But the boundary layer feels this static pressure and will respond to it.

Over the front surface of the airfoil up to the shoulder, an assisting favourable pressure gradient exists (pressure decreasing with distance downstream). The airflow speeds up along the airfoil. The flow is laminar and a laminar boundary layer is present. This laminar boundary layer grows in thickness along the airfoil. When the shoulder is reached, however, the air molecules are moving slower than in the ideal fluid case. This is an unfavourable condition because the previous ideal flow just came to rest at the trailing edge of the airfoil. It would appear now, with viscosity present, that the flow will come to rest at some distance before the trailing edge is reached.

As the airflow moves from the shoulder to the rear surface of the airfoil, the static-pressure gradient is unfavourable (increasing pressure with downstream distance). The air molecules must push against both this unfavourable pressure gradient and the viscous forces. At the transition point, the character of the airflow changes and the laminar boundary layer quickly becomes a turbulent boundary layer. This turbulent boundary layer continues to thicken downstream. Pushing against an unfavourable pressure gradient and viscosity is too much for the airflow, and at some point, the airflow stops completely. The boundary layer has stalled short of reaching the trailing edge. (Remember that the airflow reached the trailing edge before stopping in the ideal fluid case).

This stall point is known as the separation point. All along a line starting from this point outward into the airflow, the airflow is stalling. Beyond this line, the airflow is actually moving backward, upstream toward the nose before turning around. This is a region of eddies and whirlpools and represents "dead" air that is disrupting the flow field away from the airfoil. Thus, the airflow outside the dead air region is forced to flow away and around it. The region of eddies is called the wake behind the airfoil.

Up to the separation point, the difference between the static-pressure distribution for ideal fluid flow and real airflow is not very large but once separation occurs, the pressure field in greatly modified. In the ideal fluid case, the net static-pressure force acting on the front surface of the airfoil (up to the shoulder) parallel to the free stream exactly opposed and cancelled that acting on the rear surfaces of the airfoil. Under real airflow conditions, however, this symmetry and cancellation of forces is destroyed. The net static-pressure force acting on the front surface parallel to the free-stream direction now exceeds that acting on the rear surface. The net result is a drag force due to the asymmetric pressure distribution called pressure drag. This is a drag in addition to the skin-friction drag due to the shearing forces (internal friction) in the boundary layer. Additionally, the modification of the static-pressure distribution causes a decrease in the pressure lift from the ideal fluid case. The effect of viscosity is that the lift is reduced and a total drag composed of skin-friction drag and pressure drag is present. Both of these are detrimental effects.

It should be emphasized that similar processes are occurring on all the components of the aircraft to one degree or another, not only the airfoil.

Thus, the effects of a real fluid flow are the result of the viscosity of the fluid. The viscosity causes a boundary layer and, hence, a skin-friction drag. The flow field is disrupted because of viscosity to the extent that a pressure drag arises. Also, the net pressure lift is reduced.

## Interference Drag

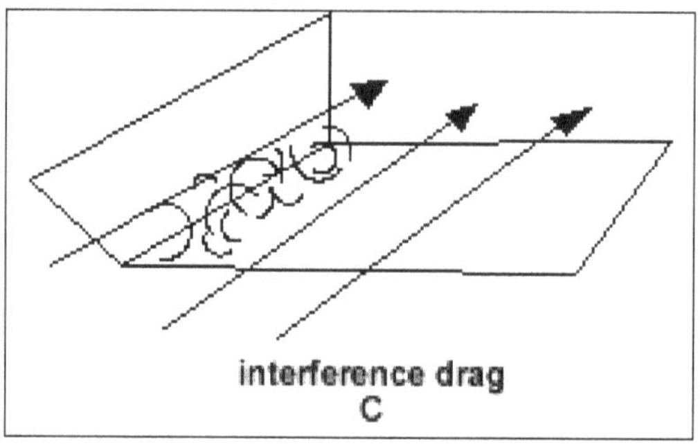


Surfaces at angles to each other as in (C) create turbulence in the region of the joint. This occurs most frequently at the intersection of the fuselage and wing.

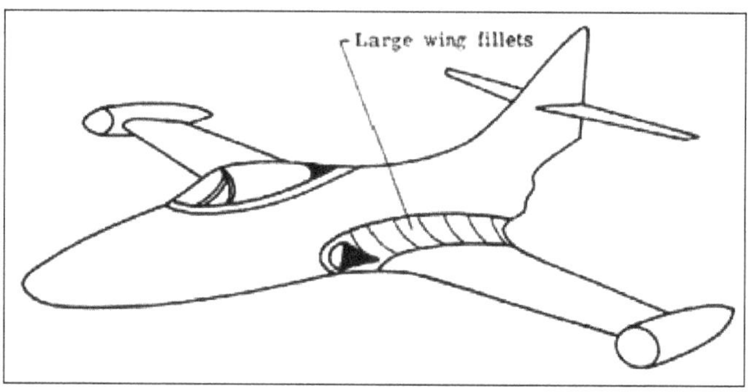


This figure shows a Grumman F9F Panther Jet with a large degree of filleting to reduce drag.

## Lift versus Drag

An aircraft with a given total gross weight can be operated in level flight over a range of power settings and airspeeds. Since Lift and Weight must be equal in order to maintain level flight, it is obvious that there is a relationship between Lift (L), Airspeed (V), and

Angle of Attack (AT). This relationship can be "generalized" with the following expression.

Lift = Angle of Attack x Velocity

Since angle of attack and speed also have a relationship to Induced Drag and Parasite Drag, the relationship of Lift/Drag is shown by the graph below.

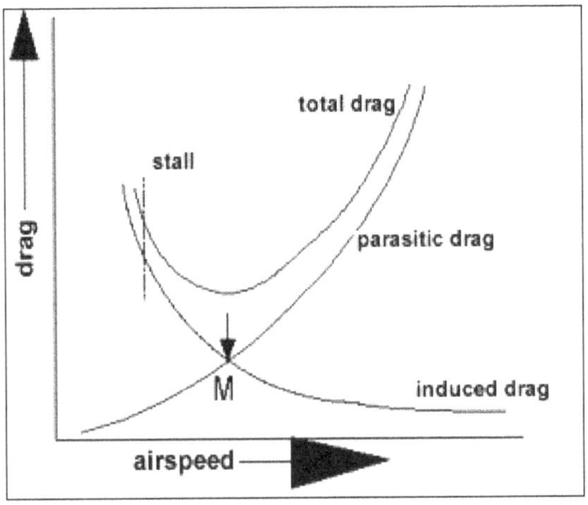


Parasite drag increases with speed. Induced drag decreases with speed. The SUM of the two drags (Total Drag curve) shows that there is only one airspeed for a given airplane and load that provides MINIMUM total drag. This is the point M which is the maximum lift over drag ratio (L/D). It is the airspeed at which the aircraft can glide the farthest without power (maximum glide range). This is the airspeed which should immediately be set up in the event of a power failure. This maximum glide airspeed is different for each aircraft design.

## References

- What-is-aerodynamics-47930: livescience.com, Retrieved 17 August, 2020
- Anderson, John (2006). Hypersonic and High-Temperature Gas Dynamics (Second ed.). AIAA Education Series. ISBN 1-56347-780-7
- Kurganov, V.A. (3 February 2011), Adiabatic Wall Temperature, Thermopedia, doi:10.1615/AtoZ.a.adiabatic_wall_temperature, retrieved 2015-10-03
- Angle-of-attack: aviation-history.com, Retrieved 18 June, 2020
- Anderson, John D. (2005). Introduction to flight (5. ed., internat. ed.). Boston [u.a.]: McGraw-Hill. pp. 274–275. ISBN 9780071238182
- Drag, training-flight-training-aero: pilotfriend.com, Retrieved 26 July, 2020
- Schmitz, Sven (2019). Aerodynamics of wind turbines: a physical basis for analysis and design. Hoboken: Wiley. p. 35. ISBN 9781119405610

# 3

# Structure and Design of Aircrafts

The aircraft structure and design process is dependent on certain factors such as customer and manufacturer demand, safety protocols, and physical and economic constraints. The structure of aircrafts consists of a fuselage, wings, an empennage, landing gear, propellers and a powerplant. This chapter discusses aircraft structure and design in detail.

## AIRCRAFT BASIC CONSTRUCTION

The airframe of a fixed-wing aircraft consists of the following five major units:

- Fuselage.
- Wings.
- Stabilizers.
- Flight controls surfaces.
- Landing gear.

A rotary-wing aircraft consists of the following four major units:

- Fuselage.
- Landing gear.
- Main rotor assembly.
- Tail rotor assembly.

You need to be familiar with the terms used for aircraft construction to work in an aviation rating.

### Structural Stress

The primary factors to consider in aircraft structures are strength, weight, and reliability. These factors determine the requirements to be met by any material used to construct or repair the aircraft. Airframes must be strong and light in weight. An aircraft built so heavy that it couldn't support more than a few hundred pounds of additional weight would be useless. All materials used to construct an aircraft must be reliable. Reliability minimizes the possibility of dangerous and unexpected failures.

Many forces and structural stresses act on an aircraft when it is flying and when it is static. When it is static, the force of gravity produces weight, which is supported by the landing gear. The landing gear absorbs the forces imposed on the aircraft by takeoffs and landings.

During flight, any maneuver that causes acceleration or deceleration increases the forces and stresses on the wings and fuselage.

Stresses on the wings, fuselage, and landing gear of aircraft are tension, compression, shear, bending, and torsion. These stresses are absorbed by each component of the wing structure and transmitted to the fuselage structure. The empennage (tail section) absorbs the same stresses and transmits them to the fuselage. These stresses are known as loads, and the study of loads is called a stress analysis. Stresses are analyzed and considered when an aircraft is designed. The stresses acting on an aircraft are shown in figure.

## Tension

Tension is defined as pull. It is the stress of stretching an object or pulling at its ends. Tension is the resistance to pulling apart or stretching produced by two forces pulling in opposite directions along the same straight line. For example, an elevator control cable is in additional tension when the pilot moves the control column.

## Compression

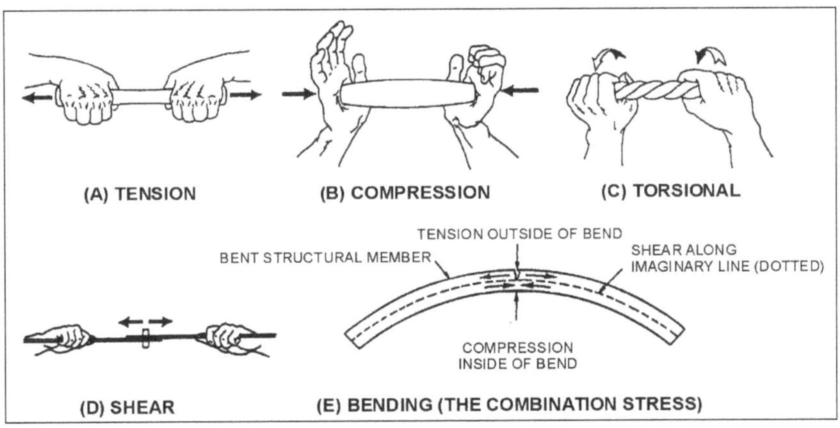


Five stresses acting on an aircraft.

If forces acting on an aircraft move toward each other to squeeze the material, the stress is called compression. Compression is the opposite of tension. Tension is pull, and compression is push. Compression is the resistance to crushing produced by two forces pushing toward each other in the same straight line. For example, when an airplane is on the ground, the landing gear struts are under a constant compression stress.

## Shear

Cutting a piece of paper with scissors is an example of a shearing action. In an aircraft structure, shear is a stress exerted when two pieces of fastened material tend to separate. Shear stress is the outcome of sliding one part over the other in opposite directions. The rivets and bolts of an aircraft experience both shear and tension stresses.

## Bending

Bending is a combination of tension and compression. For example, when bending a piece of tubing, the upper portion stretches (tension) and the lower portion crushes together (compression). The wing spars of an aircraft in flight are subject to bending stresses.

## Torsion

Torsional stresses result from a twisting force. When you wring out a chamois skin, you are putting it under torsion. Torsion is produced in an engine crankshaft while the engine is running. Forces that produce torsional stress also produce torque.

## Varying Stress

All structural members of an aircraft are subject to one or more stresses. Sometimes a structural member has alternate stresses; for example, it is under compression one instant and under tension the next. The strength of aircraft materials must be great enough to withstand maximum force of varying stresses.

## Specific Action of Stresses

Engine torque creates torsion stress in aircraft fuselages.

You need to understand the stresses encountered on the main parts of an aircraft. A knowledge of the basic stresses on aircraft structures will help you understand why aircraft are built the way they are. The fuselage of an aircraft is subject the fives types of stress—torsion, bending, tension, shear, and compression.

Torsional stress in a fuselage is created in several ways. For example, torsional stress is encountered in engine torque on turboprop aircraft. Engine torque tends to rotate the aircraft in the direction opposite to the direction the propeller is turning. This force creates a torsional stress in the fuselage. Figure shows the effect of the rotating propellers. Also, torsional stress on the fuselage is created by the action of the ailerons when the aircraft is maneuvered.

When an aircraft is on the ground, there is a bending force on the fuselage. This force occurs because of the weight of the aircraft. Bending increases when the aircraft makes a carrier landing. This bending action creates a tension stress on the lower skin of the fuselage and a compression stress on the top skin. Bending action is shown in figure. These stresses are transmitted to the fuselage when the aircraft is in flight. Bending occurs because of the reaction of the airflow against the wings and empennage. When the aircraft is in flight, lift forces act upward against the wings, tending to bend them upward. The wings are prevented from folding over the fuselage by the resisting strength of the wing structure. The bending action creates a tension stress on the bottom of the wings and a compression stress on the top of the wings.

## Construction Materials

An aircraft must be constructed of materials that are both light and strong. Early aircraft were made of wood. Lightweight metal alloys with a strength greater than wood were developed and used on later aircraft. Materials currently used in aircraft construction are classified as either metallic materials or nonmetallic materials.

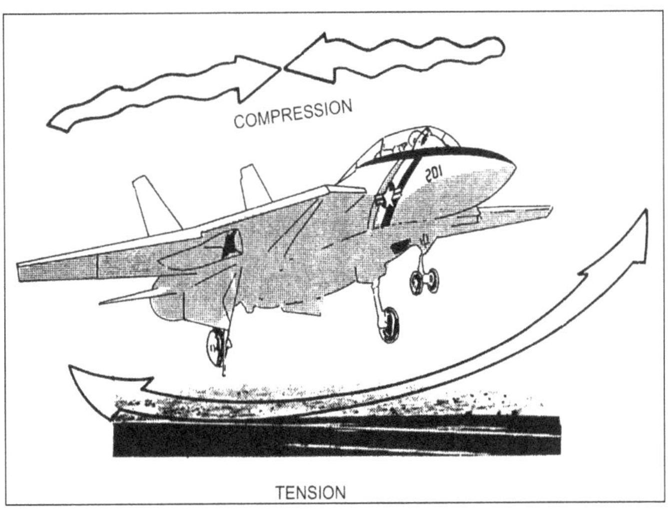


Bending action occurring during carrier landing.

## Metallic Materials

The most common metals used in aircraft construction are aluminum, magnesium, titanium, steel, and their alloys.

## Alloys

An alloy is composed of two or more metals. The metal present in the alloy in the largest amount is called the base metal. All other metals added to the base metal are called alloying elements. Adding the alloying elements may result in a change in the properties of the base metal. For example, pure aluminum is relatively soft and weak. However, adding small amounts or copper, manganese, and magnesium will increase aluminum's strength many times. Heat treatment can increase or decrease an alloy's strength and hardness. Alloys are important to the aircraft industry. They provide materials with properties that pure metals do not possess.

## Aluminum

Aluminum alloys are widely used in modern aircraft construction. Aluminum alloys are valuable because they have a high strength-to-weight ratio. Aluminum alloys are corrosion resistant and comparatively easy to fabricate. The outstanding characteristic of aluminum is its lightweight.

## Magnesium

Magnesium is the world's lightest structural metal. It is a silvery-white material that weighs two-thirds as much as aluminum. Magnesium is used to make helicopters. Magnesium's low resistance to corrosion has limited its use in conventional aircraft.

## Titanium

Titanium is a lightweight, strong, corrosionresistant metal. Recent developments make titanium ideal for applications where aluminum alloys are too weak and stainless steel is too heavy. Additionally, titanium is unaffected by long exposure to seawater and marine atmosphere.

## Steel Alloys

Alloy steels used in aircraft construction have great strength, more so than other fields of engineering would require. These materials must withstand the forces that occur on today's modern aircraft. These steels contain small percentages of carbon, nickel, chromium, vanadium, and molybdenum. High-tensile steels will stand stress of 50 to 150 tons per square inch without failing. Such steels are made into tubes, rods, and wires.

Another type of steel used extensively is stainless steel. Stainless steel resists corrosion and is particularly valuable for use in or near water.

## Nonmetallic Materials

In addition to metals, various types of plastic materials are found in aircraft construction. Some of these plastics include transparent plastic, reinforced plastic, composite, and carbon-fiber materials.

## Transparent Plastic

Transparent plastic is used in canopies, windshields, and other transparent enclosures. You need to handle transparent plastic surfaces carefully because they are relatively soft and scratch easily. At approximately 225 °F, transparent plastic becomes soft and pliable.

## Reinforced Plastic

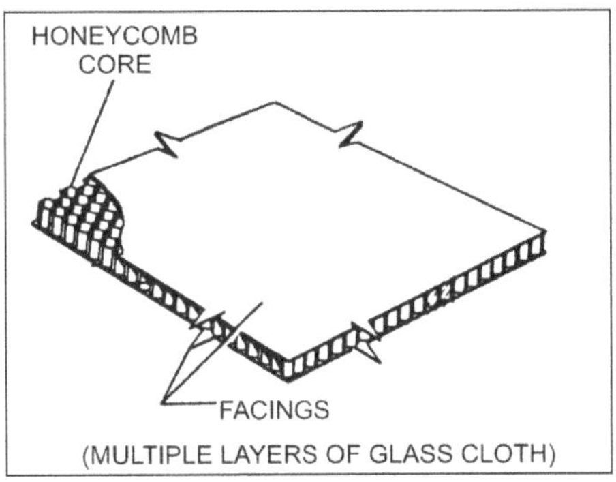


Reinforced plastic.

Reinforced plastic is used in the construction of radomes, wingtips, stabilizer tips, antenna covers, and flight controls. Reinforced plastic has a high strength-to-weight ratio and is resistant to mildew and rot. Because it is easy to fabricate, it is equally suitable for other parts of the aircraft.

Reinforced plastic is a sandwich-type material. It is made up of two outer facings and a center layer. The facings are made up of several layers of glass cloth, bonded together with a liquid resin. The core material (center layer) consists of a honeycomb structure made of glass cloth. Reinforced plastic is fabricated into a variety of cell sizes.

## Composite and Carbon Fiber Materials

High-performance aircraft require an extra high strength-to-weight ratio material. Fabrication of composite materials satisfies this special requirement. Composite materials are constructed by using several layers of bonding materials (graphite epoxy or boron epoxy). These materials are mechanically fastened to conventional substructures. Another type of composite construction consists of thin graphite epoxy skins bonded to an aluminum honeycomb core. Carbon fiber is extremely strong, thin fiber made by heating synthetic fibers, such as rayon, until charred, and then layering in cross sections.

## Fixed-wing Aircraft

The principal structural units of a fixed-wing aircraft are the fuselage, wings, stabilizers, flight control surfaces, and landing gear. Figure shows these units of a naval aircraft.

The terms left or right used in relation to any of the structural units refer to the right or left hand of the pilot seated in the cockpit.

## Fuselage

The fuselage is the main structure, or body, of the aircraft. It provides space for personnel, cargo, controls, and most of the accessories. The power plant, wings, stabilizers, and landing gear are attached to it.

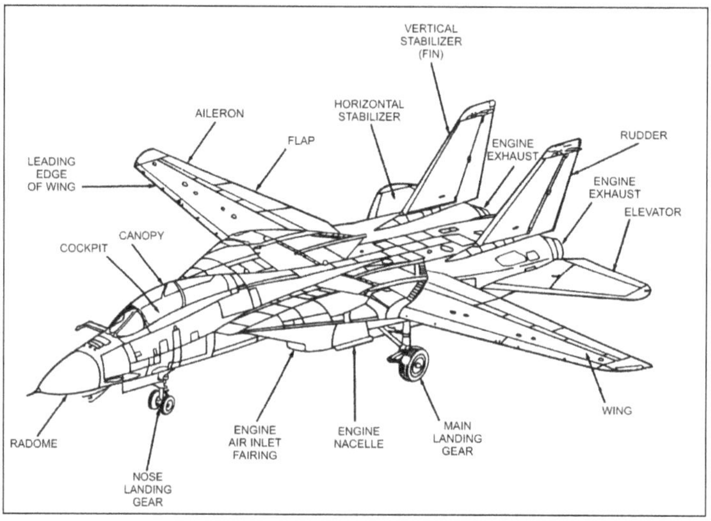


Principal structural units on an F-14 aircraft.

There are two general types of fuselage construction—welded steel truss and monocoque designs. The welded steel truss was used in smaller Navy aircraft, and it is still being used in some helicopters.

The monocoque design relies largely on the strength of the skin, or covering, to carry various loads. The monocoque design may be divided into three classes—monocoque, semimonocoque, and reinforced shell.

- The true monocoque construction uses formers, frame assemblies, and bulkheads to give shape to the fuselage. However, the skin carries the primary stresses. Since no bracing members are present, the skin must be strong enough to keep the fuselage rigid. The biggest problem in monocoque construction is maintaining enough strength while keeping the weight within limits.

- Semimonocoque design overcomes the strength-to-weight problem of monocoque construction. In addition to having formers, frame assemblies, and bulkheads, the semimonocoque construction has the skin reinforced by longitudinal members.
- The reinforced shell has the skin reinforced by a complete framework of structural members. Different portions of the same fuselage may belong to any one of the three classes. Most are considered to be of semimonocoque-type construction.

The semimonocoque fuselage is constructed primarily of aluminum alloy, although steel and titanium are found in high-temperature areas. Primary bending loads are taken by the longerons, which usually extend across several points of support. The longerons are supplemented by other longitudinal members known as stringers. Stringers are more numerous and lightweight than longerons.

The vertical structural members are referred to as bulkheads, frames, and formers. The heavier vertical members are located at intervals to allow for concentrated loads. These members are also found at points where fittings are used to attach other units, such as the wings and stabilizers.

The stringers are smaller and lighter than longerons and serve as fill-ins. They have some rigidity but are chiefly used for giving shape and for attachment of skin. The strong, heavy longerons hold the bulkheads and formers. The bulkheads and formers hold the stringers. All of these join together to form a rigid fuselage framework. Stringers and longerons prevent tension and compression stresses from bending the fuselage.

The skin is attached to the longerons, bulkheads, and other structural members and carries part of the load. The fuselage skin thickness varies with the load carried and the stresses sustained at particular location.

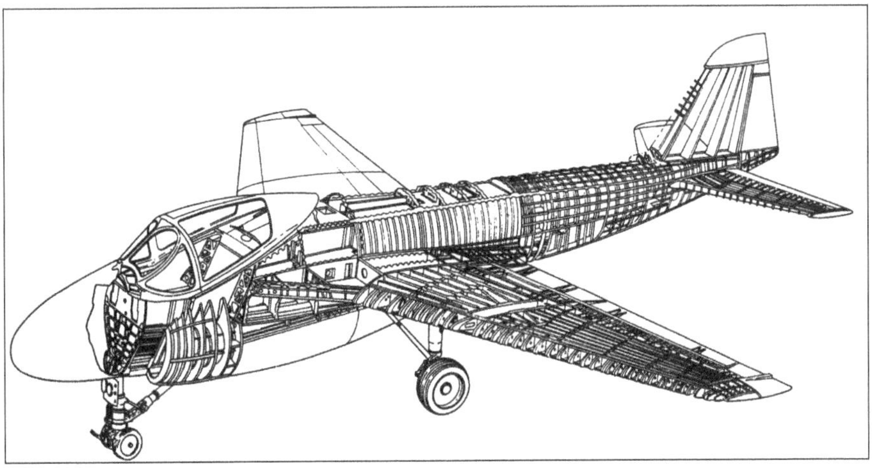

Semimonocoque fuselage construction.

There are a number of advantages in using the semimonocoque fuselage:

- The bulkhead, frames, stringers, and longerons aid in the design and construction of a streamlined fuselage. They add to the strength and rigidity of the structure.
- The main advantage of the semimonocoque construction is that it depends on many structural members for strength and rigidity. Because of its stressed skin construction, a semimonocoque fuselage can withstand damage and still be strong enough to hold together.

Points on the fuselage are located by station numbers. Station 0 is usually located at or near the nose of the aircraft. The other stations are located at measured distances (in inches) aft of station 0. A typical station diagram is shown in figure. On this particular aircraft, fuselage station (FS) 0 is located 93.0 inches forward of the nose.

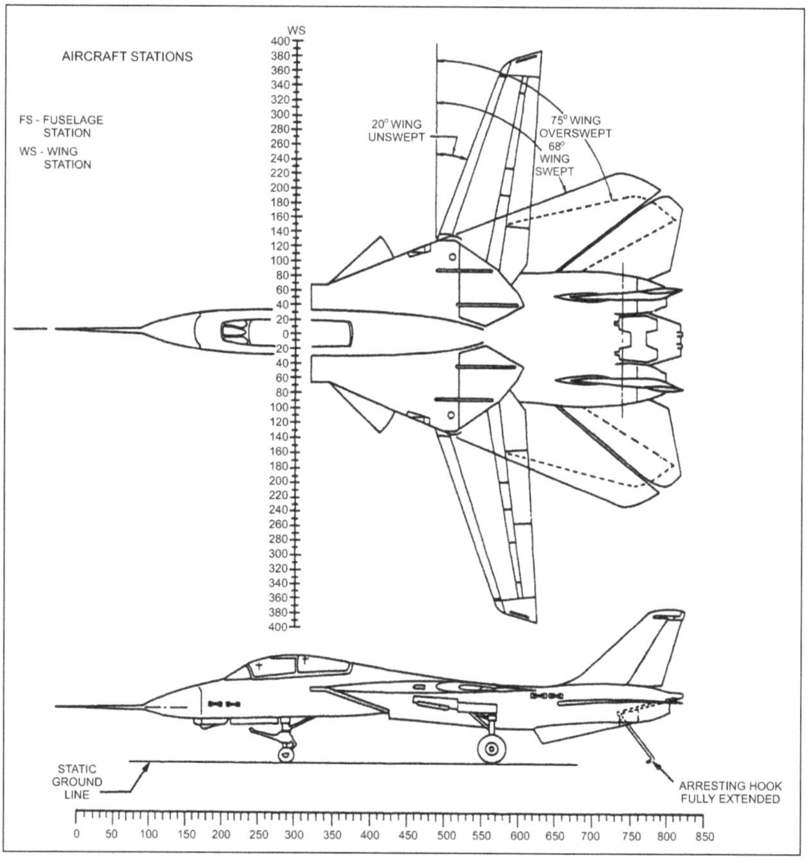


Fuselage station diagram of an F-14 aircraft.

## Wings

Wings develop the major portion of the lift of a heavier-than-air aircraft. Wing structures carry some of the heavier loads found in the aircraft structure. The particular design of a wing depends on many factors, such as the size, weight, speed, rate of climb, and use of the aircraft. The wing must be constructed so that it holds its aerodynamics shape under the extreme stresses of combat maneuvers or wing loading.

Wing construction is similar in most modern aircraft. In its simplest form, the wing is a framework made up of spars and ribs and covered with metal. The construction of an aircraft wing is shown in figure.

Spars are the main structural members of the wing. They extend from the fuselage to the tip of the wing. All the load carried by the wing is taken up by the spars. The spars are designed to have great bending strength. Ribs give the wing section its shape, and they transmit the air load from the wing covering to the spars. Ribs extend from the leading edge to the trailing edge of the wing.

In addition to the main spars, some wings have a false spar to support the ailerons and flaps. Most aircraft wings have a removable tip, which streamlines the outer end of the wing.

Most Navy aircraft are designed with a wing referred to as a wet wing. This term describes the wing that is constructed so it can be used as a fuel cell. The wet wing is sealed with a fuel-resistant compound as it is built. The wing holds fuel without the usual rubber cells or tanks.

The wings of most naval aircraft are of all metal, full cantilever construction. Often, they may be folded for carrier use. A full cantilever wing structure is very strong. The wing can be fastened to the fuselage without the use of external bracing, such as wires or struts.

A complete wing assembly consists of the surface providing lift for the support of the aircraft. It also provides the necessary flight control surfaces.

The flight control surfaces on a simple wing may include only ailerons and trailing edge flaps. The more complex aircraft may have a variety of devices, such as leading edge flaps, slats, spoilers, and speed brakes.

Various points on the wing are located by wing station numbers. Wing station (WS) 0 is located at the centerline of the fuselage, and all wing stations are measured (right or left) from this point (in inches).

## Stabilizers

The stabilizing surfaces of an aircraft consist of vertical and horizontal airfoils. They are called the vertical stabilizer (or fin) and horizontal stabilizer. These two airfoils, along with the rudder and elevators, form the tail section. For inspection and maintenance purposes, the entire tail section is considered a single unit called the empennage.

The main purpose of stabilizers is to keep the aircraft in straight-and-level flight. The vertical stabilizer maintains the stability of he aircraft about its vertical axis. This is known as directional stability. The vertical stabilizer usually serves as the base to which the rudder is attached. The horizontal stabilizer provides stability of the aircraft about its lateral axis. This is known as longitudinal stability. The horizontal stabilizer usually serves as the base to which the elevators are attached. On many newer, high-performance aircraft, the entire vertical and/or horizontal stabilizer is a movable airfoil. Without the movable airfoil, the flight control surfaces would lose their effectiveness at extremely high altitudes.

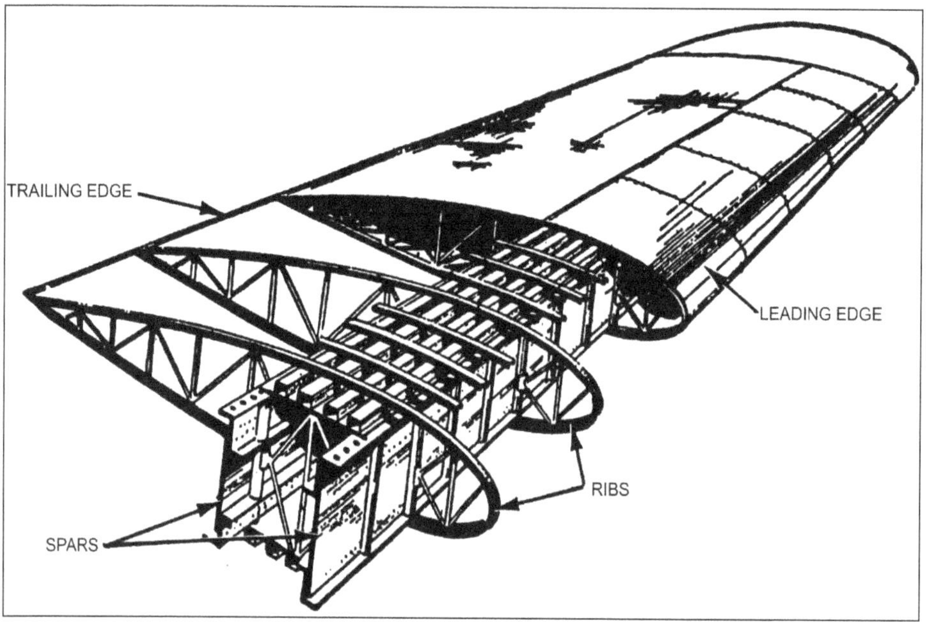


Stabilizer construction is similar to wing construction. For greater strength, especially in the thinner airfoil sections typical of trailing edges, a honeycomb-type construction is used. Some larger carrier-type aircraft have vertical stabilizers that are folded hydraulically to aid aircraft movement aboard aircraft carriers.

## Flight Control Surfaces

Flight control surfaces are hinged (movable) airfoils designed to change the attitude of the aircraft during flight. These surfaces are divided into three groups—primary, secondary, and auxiliary.

## Primary Group

The primary group of flight control surfaces includes ailerons, elevators, and rudders. The ailerons attach to the trailing edge of the wings. They control the rolling (or banking) motion of the aircraft. This action is known as longitudinal control.

The elevators are attached to the horizontal stabilizer and control the climb or descent (pitching motion) of the aircraft. This action is known as lateral control.

The rudder is attached to the vertical stabilizer. It determines the horizontal flight (turning or yawing motion) of the aircraft. This action is known as directional control.

The ailerons and elevators are operated from the cockpit by a control stick on single-engine aircraft. A yoke and wheel assembly operates the ailerons and elevators on multiengine aircraft, such as transport and patrol aircraft. The rudder is operated by foot pedals on all types of aircraft.

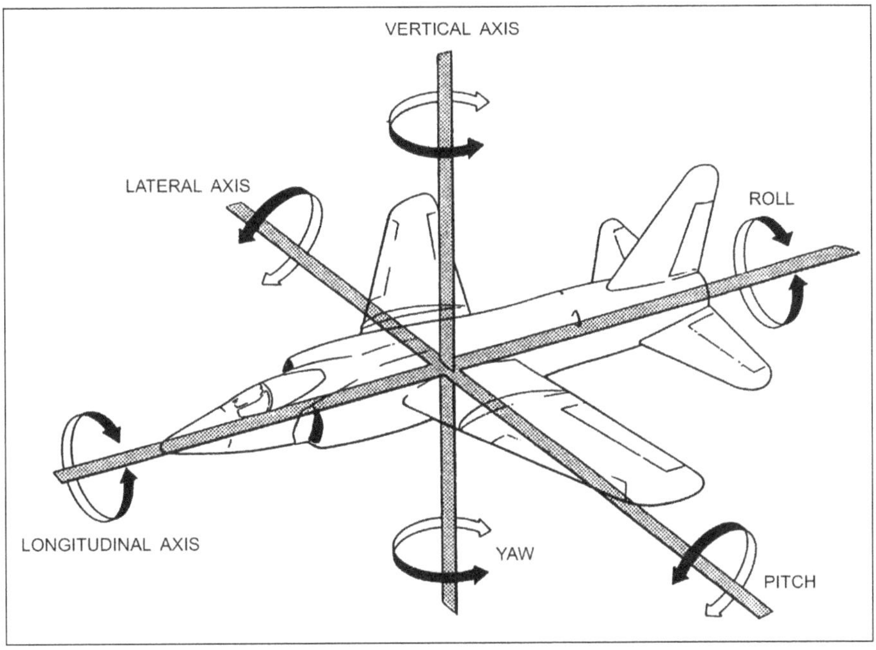


Axes and fundamental movements of the aircraft.

## Secondary Group

The secondary group includes the trim tabs and spring tabs. Trim tabs are small airfoils recessed into the trailing edges of the primary control surface. Each trim tab hinges to its parent primary control surface, but operates by an independent control. Trim tabs let the pilot trim out an unbalanced condition without exerting pressure on the primary controls.

Spring tabs are similar in appearance to trim tabs but serve an entirely different purpose. Spring tabs are used for the same purpose as hydraulic actuators. They aid the pilot in moving a larger control surface, such as the ailerons and elevators.

## Auxiliary Group

The auxiliary group includes the wing flaps, spoilers, speed brakes, and slats.

Wing flaps: Wing flaps give the aircraft extra lift. Their purpose is to reduce the landing speed. Reducing the landing speed shortens the length of the landing rollout. Flaps help the pilot land in small or obstructed areas by increasing the glide angle without greatly increasing the approach speed. The use of flaps during takeoff serves to reduce the length of the takeoff run.

Some flaps hinge to the lower trailing edges of the wings inboard of the ailerons. Leading edge flaps are used on the F-14 Tomcat and F/A-18 Hornet. Four types of flaps are shown in figure. The plain flap forms the trailing edge of the airfoil when the flap is in the up position. In the split flap, the trailing edge of the airfoil is split, and the lower half is hinged and lowers to form the flap. The fowler flap operates on rollers and tracks, causing the lower surface of the wing to roll out and then extend downward. The leading edge flap operates like the plain flap. It is hinged on the bottom side. When actuated, the leading edge of the wing actually extends in a downward direction to increase the camber of the wing. Landing flaps are used in conjunction with other types of flaps.

Spoilers: Spoilers are used to decrease wing lift. The specific design, function, and use vary with different aircraft. On some aircraft, the spoilers are long narrow surfaces, hinged at their leading edge to the upper surfaces of the wings. In the retracted position, they are flush with the wing skin. In the raised position, they greatly reduce wing lift by destroying the smooth flow of air over the wing surface.

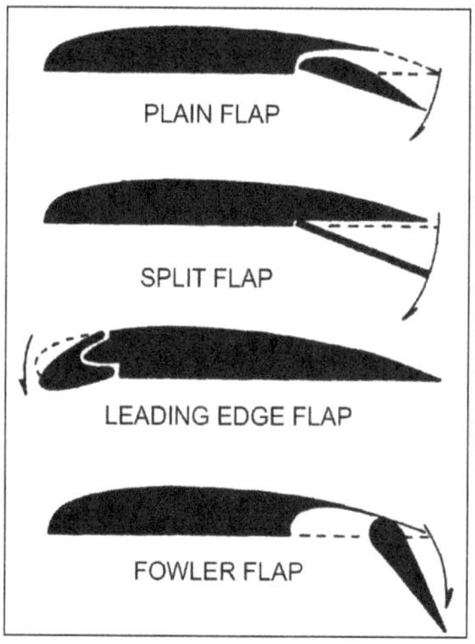


Types of flaps.

Speed brakes: Speed brakes are movable control surfaces used for reducing the speed of the aircraft. Some manufacturers refer to them as dive brakes; others refer to them as dive flaps. On some aircraft, they're hinged to the sides or bottom of the fuselage. Regardless of their location, speed brakes serve the same purpose—to keep the airspeed from building too high when the aircraft dives. Speed brakes slow the aircraft's speed before it lands.

Slats: Slats are movable control surfaces that attach to the leading edge of the wing. When the slat is retracted, it forms the leading edge of the wing. When the slat is open (extended forward), a slot is created between the slat and the wing leading edge. High-energy air is introduced into the boundary layer over the top of the wing. At low airspeeds, this action improves the lateral control handling characteristics. This allows the aircraft to be controlled at airspeeds below normal landing speed. The high-energy air that flows over the top of the wing is known as boundary layer control air. Boundary layer control is intended primarily for use during operations from carriers. Boundary layer control air aids in catapult takeoffs and arrested landings. Boundary control air can also be accomplished by directing high-pressure engine bleed air across the top of the wing or flap surface.

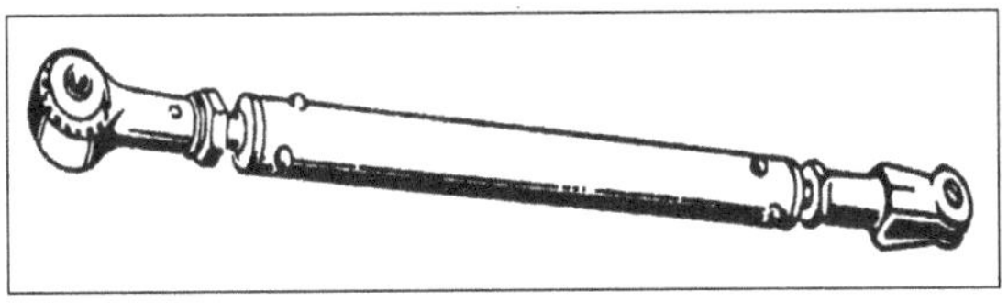

Push-pull tube assembly.

## Flight Control Mechanisms

The term flight control refers to the linkage that connects the control(s) in the cockpit with the flight control surfaces. There are several types of flight controls in naval aircraft; some are manually operated while others are power operated.

Manually operated flight control mechanisms are further divided into three groups—cable operated, push-pull tube operated, and torque tube operated. Some systems may combine two or more of these types.

In the manually operated cable system, cables are connected from the control in the cockpit to a bell crank or sector. The bell crank is connected to the control surface. Movement of the cockpit controls transfers force through the cable to the bell crank, which moves the control surface.

In a push-pull tube system, metal push-pull tubes (or rods) are used as a substitute for the cables. Push-pull tubes get their name from the way they transmit force.

In the torque tube system, metal tubes (rods) with gears at the ends of the tubes are used. Motion is transmitted by rotating the tubes and gears.

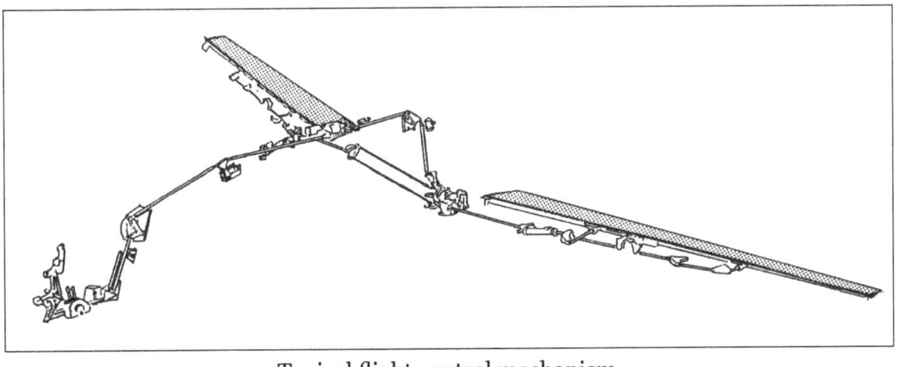

Typical flight control mechanism.

On all high-performance aircraft, the control surfaces have great pressure exerted on them. At high airspeed, it is physically impossible for the pilot to move the controls manually. As a result, power-operated control mechanisms are used. In a power-operated system, a hydraulic actuator (cylinder) is located within the linkage to assist the pilot in moving the control surface.

A typical flight control mechanism is shown in figure. This is the elevator control of a lightweight trainer-type aircraft. It consists of a combination of push-pull tubes and cables.

The control sticks in the system shown in figure are connected to the forward sector by push-pull tubes. The forward sector is connected to the aft (rear) sector by means of cable assemblies. The aft sector is connected to the flight control by another push-pull tube assembly.

## Landing Gear

Before World War II, aircraft were made with their main landing gear located behind the center of gravity. An auxiliary gear under the fuselage nose was added. This arrangement became known as the tricycle type of landing gear. Nearly all present-day Navy aircraft are equipped with tricycle landing gear. The tricycle gear has the following advantages over older landing gear:

- More stable in motion on the ground.
- Maintains the fuselage in a level position.
- Increases the pilot's visibility and control.
- Makes landing easier, especially in cross winds.

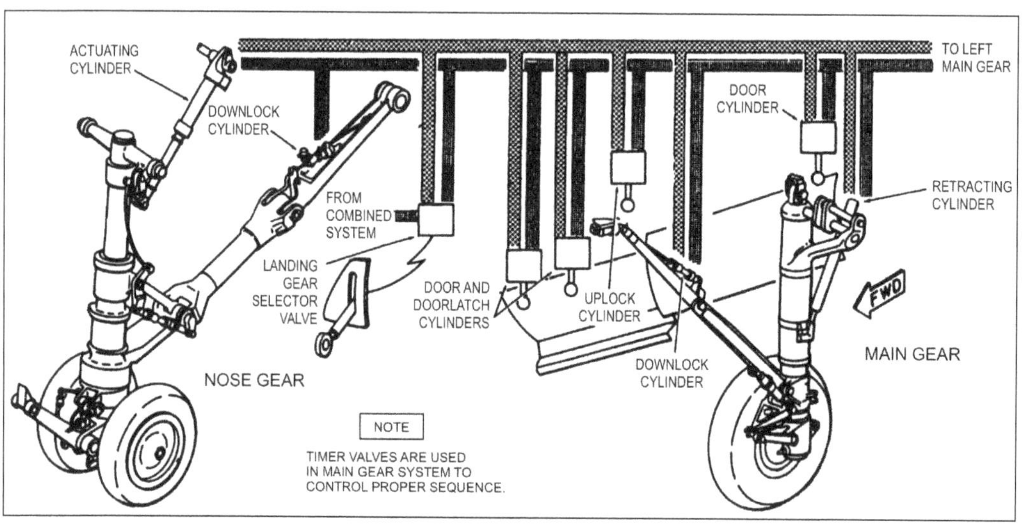


Typical landing gear system.

The landing gear system consists of three retractable landing gear assemblies. Each main landing gear has a conventional air-oil shock strut, a wheel brake assembly, and a wheel and tire assembly. The nose landing gear has a conventional air-oil shock strut, a shimmy damper, and a wheel and tire assembly.

The shock strut is designed to absorb the shock that would otherwise be transmitted to the airframe during landing, taxiing, and takeoff. The air-oil strut is used on all naval aircraft. This type of strut has two telescoping cylinders filled with hydraulic fluid and compressed air or nitrogen. Figure shows the internal construction of one type of air-oil shock strut.

The main landing gear is equipped with brakes for stopping the aircraft and assisting the pilot in steering the aircraft on the ground.

The nose gear of most aircraft can be steered from the cockpit. This provides greater ease and safety on the runway when landing and taking off and on the taxiway in taxiing.

## Arresting Gear

A carrier-type aircraft is equipped with an arresting hook for stopping the aircraft when it lands on the carrier. The arresting gear has an extendible hook and the mechanical, hydraulic, and pneumatic equipment necessary for hook operation.The arresting hook on most aircraft releases mechanically, lowers pneumatically, and raises hydraulically.

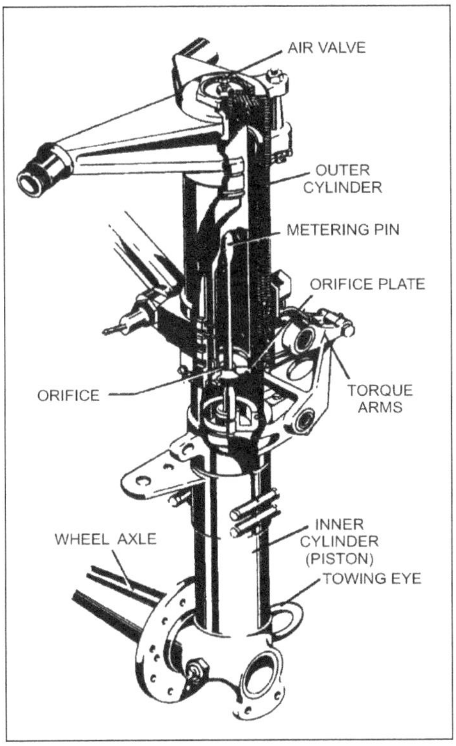


Internal construction of a shock strut.

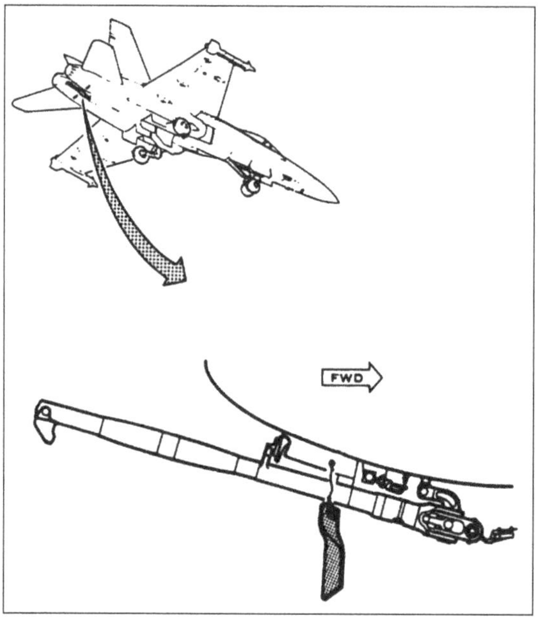


Arresting gear installation.

The hook hinges from the structure under the rear of the aircraft. A snubber meters hydraulic fluid and works in conjunction with nitrogen pressure. The snubber holds the hook down and prevents it from bouncing when it strikes the carrier deck.

## Catapult Equipment

Carrier aircraft have built-in equipment for catapulting off the aircraft carrier. Older aircraft had hooks on the airframe that attached to the cable bridle. The bridle hooks the aircraft to the ship's catapult. Newer aircraft have a launch bar built into the nose landing gear assembly. The holdback assembly allows the aircraft to be secured to the carrier deck for full-power turnup of the engine prior to takeoff. For nose gear equipment, a track attaches to the deck to guide the nosewheel into position. The track has provisions for attaching the nose gear to the catapult shuttle and for holdback.

The holdback tension bar separates when the catapult is fired, allowing the aircraft to be launched with the engine at full power.

## Rotary-wing Aircraft

Within the past 20 years, helicopters have become a reality, and are found throughout the world. They perform countless tasks suited to their unique capabilities.

A helicopter has one or more power-driven horizontal airscrews (rotors) to develop lift and propulsion. If a single main rotor is used, it is necessary to employ a means to counteract torque. If more than one main rotor (or tandem) is used, torque is eliminated by turning each main rotor in opposite directions.

The fundamental advantage the helicopter has over fixed-wing aircraft is that lift and control are independent of forward speed. A helicopter can fly forward, backward, or sideways, or it can remain in stationary flight (hover) above the ground. No runway is required for a helicopter to take off or land. For example, the roof of an office building is an adequate landing area. The helicopter is considered a safe aircraft because the takeoff and landing speed is zero, and it has autorotational capabilities. This allows a controlled descent with rotors turning in case of engine failure in flight.

## Fuselage

Like the fuselage of a fixed-wing aircraft, the helicopter fuselage may be welded truss or some form of monocoque construction. Many Navy helicopters are of the monocoque design.

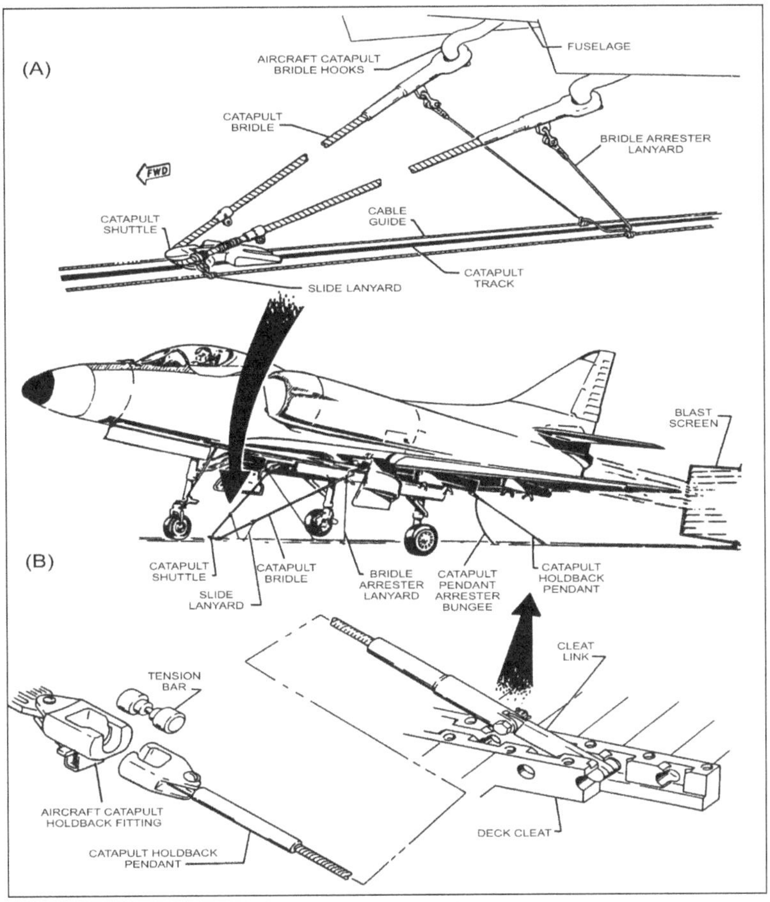


Aircraft catapult equipment.

A typical Navy helicopter, the H-60, is shown in figure. Some of its features include a single main rotor, twin engine, tractor-type canted tail rotor, controllable stabilizer, fixed landing gear, rescue hoist, external cargo hook, and weapons pylons. The fuselage consists of the entire airframe, sometimes known as the body group.

The body group is an all-metal semimonocoque construction. It consists of an aluminum and titanium skin over a reinforced aluminum frame.

## Landing Gear Group

The landing gear group includes all the equipment necessary to support the helicopter when it is not in flight. There are several types of landing gear on helicopters—conventional fixed (skid type), retractable, and nonretractable.

## Main Landing Gear

The H-60's nonretracting main landing gear consists of two single axle, air/oil type of shock-strut assemblies that mount to the fuselage. Each is equipped with tubeless tires, hydraulic disc brakes, tie-down rings, drag braces, and safety switches. They are part of the lower end of the shock strut piston.

## Tail Landing Gear

The H-60's tail landing gear is a nonretracting, dual wheel, 360-degree swiveling type. It is equipped with tubeless tires, tie-down ring, shimmy damper, tail-wheel lock, and an air/oil shock-strut, which serves as an aft touchdown point for the pilots to cushion the landing shock.

## Main Rotor Assembly

The main rotor (rotor wing) and rotor head (hub assembly) are identical in theory of flight but differ in engineering or design. They are covered here because their functions are closely related. The power plant, transmission, drive-train, hydraulic flight control, and rotor systems all work together. Neither has a function without the other.

## Rotary Wing

The main rotor on the H-60 has four identical wing blades. Other types of helicopters may have two, four, five, six, or seven blades. Figure shows some typical rotor blades.

H-60 helicopter.

Rotary-wing blades are made of titanium, aluminum alloys, fiber glass, graphite, honeycomb core, nickel, and steel. Each has a nitrogen-filled, pressurized, hollow internal spar, which runs the length of the blade. The cuff provides the attachment of the blade to the rotor hub. A titanium abrasion strip covers the entire leading edge of the spar from the cuff end to the removable blade tip faring. This extends the life of the rotor blade.

The examples shown in figure show other features—trim tabs, deicing protection, balance markings, and construction.

## Main Rotor Head/Hub Assembly

The rotor head is fully articulating and is rotated by torque from the engines through the drive train and main gearbox or transmission. The flight controls and hydraulic servos transmit movements to the rotor blades. The principal components of the rotor head are the hub and swashplate assemblies. The hub is one piece, made of titanium and sits on top of the rotor mast. Attaching components are the sleeve and spindles, blade fold components, vibration absorber, bearings, blade dampers, pitch change horns, adjustable pitch control rods, blade fold hinges, balance weights, antiflapping and droop stops, and faring.

The swashplate consists of a rotating disc (upper), stationary (lower) portion with a scissors and sleeve assembly separated by a bearing. The swashplate is permitted to slide on the main rotor vertical driveshaft and mounts on top the main transmission. The entire assembly can tilt in any direction following the motion of the flight controls.

The hydraulic servo cylinders, swashplate, and adjustable pitch control rods permit movement of the flight controls to be transmitted to the rotary-wing blades. The sleeve and spindle and blade dampers allow limited movement of the blades in relation to the hub. These movements are known as lead, lag, and flap.

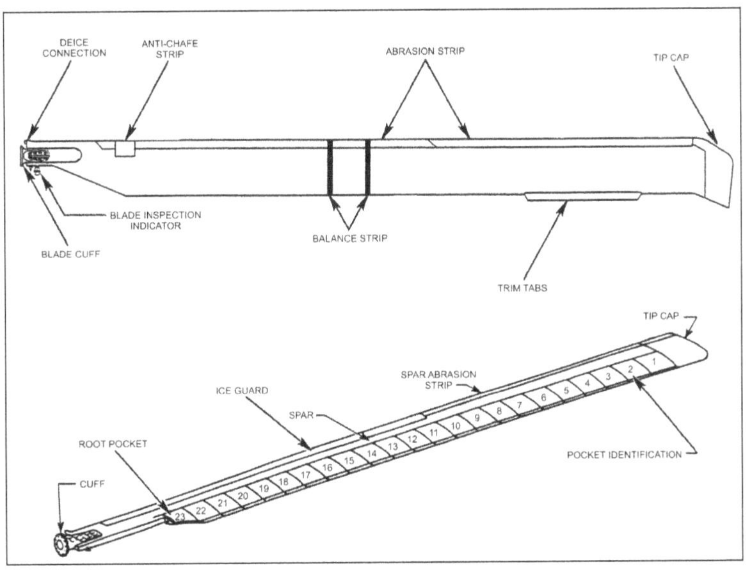


Types of main rotor blades.

- Lead occurs during slowing of the drive mechanism when the blades have a tendency to remain in motion.
- Lag is the opposite of lead and occurs during acceleration when the blade has been at rest and tends to remain at rest.
- Flap is the tendency of the blade to rise with high-lift demands as it tries to screw itself upward into the air.

Antiflapping stops and droop stops restrict flapping and conning motion of the rotary-wing head and blades at low rotor rpm when slowing or stopping.

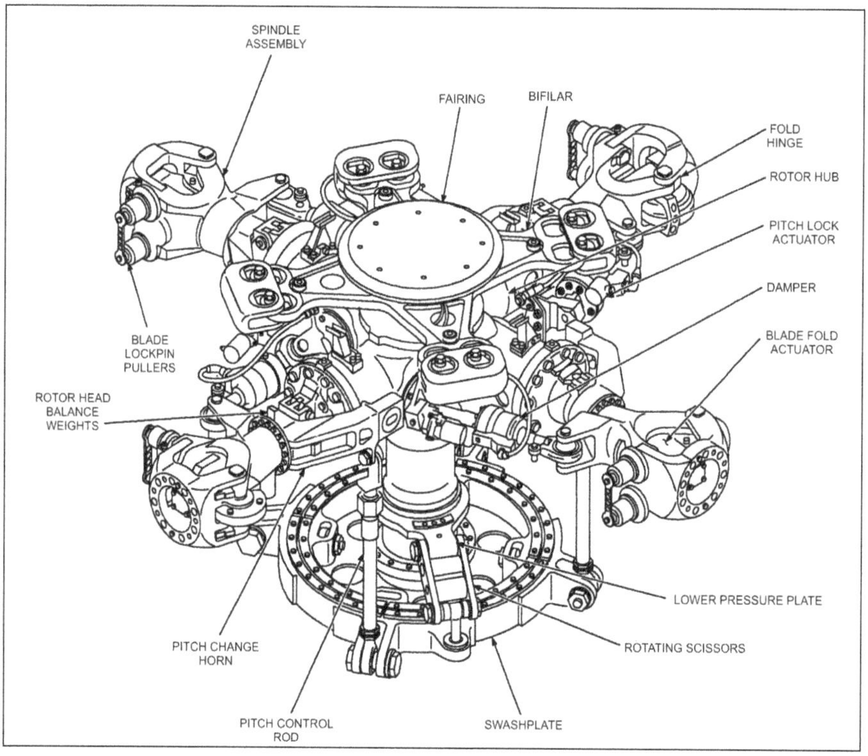


Main rotor head/hub assembly.

## Tail Rotor Group

The directional control and antitorque action of the helicopter is provided by the tail rotor group. These components are similar in function to the main rotor.

## Pylon

The pylon, shown in figure, attaches on the aircraft to the main fuselage by hinge fittings. These hinge fittings serve as the pivot point for the pylon to fold along the fuselage. Folding the pylon reduces the overall length of the helicopter, which helps for confined shipboard handling.

The pylon houses the intermediate and tail rotor gearboxes, tail rotor drive shaft, cover, tail bumper, position/anticollision lights, hydraulic servos, flight control push-pull tubes/cables/bell cranks, stabilizer/ elevator flight control surface, some antennas, and rotary rudder assembly.

## Rotary Rudder Head

The rudder head can be located on either side of the pylon, depending on the type of aircraft, and includes such items as the hub, spindle, pitch control beam, pitch change links, bearings, and tail rotor blades.

Change in blade pitch is accomplished through the pitch change shaft that moves through the horizontal shaft of the tail gearbox, which drives the rotary rudder assembly. As the shaft moves inward toward the tail gearbox, pitch of the blade is decreased. As the shaft moves outward from the tail gearbox, pitch of the blade is increased. The pitch control beam is connected by links to the forked brackets on the blade sleeves.

## Rotary Rudder Blades

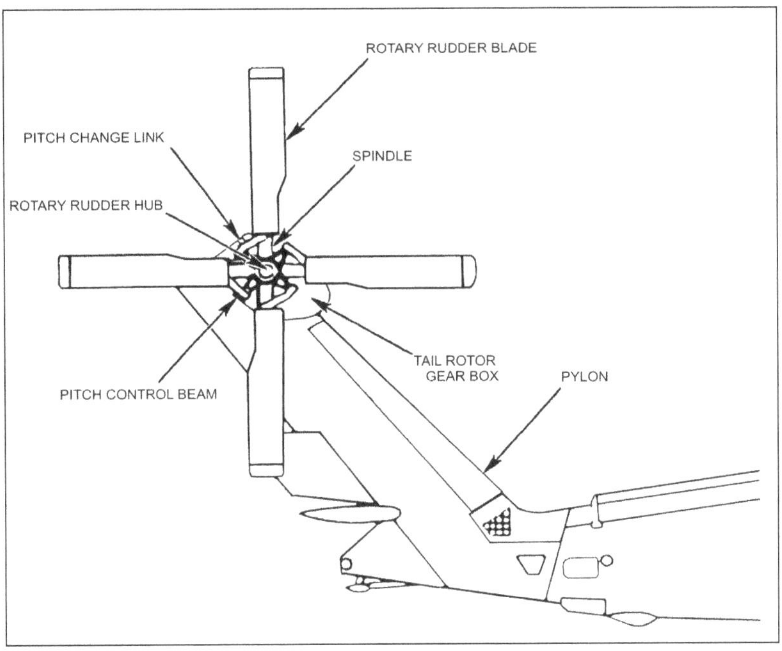


Tail rotor group.

Like the blades on a main rotor head, the blades found on a rotary rudder head may differ, depending on the type of aircraft. Tail rotor blades may consist of the following components:

- Aluminum alloy, graphite composite, or titanium spar.
- Aluminum pocket and skin with honeycomb core or cross-ply fiber glass exterior.

- Aluminum or graphite composite tip cap.
- Aluminum trailing edge cap.
- Aluminum or polyurethane and nickel abrasion leading edge strip.

Additionally, rotary rudder blades may have deicing provisions, such as electrothermal blankets that are bonded into the blade's leading edge. or a neoprene anti-icing guard embedded with electrical heating elements.

## Aircraft Hydraulic Systems

The aircraft hydraulic systems found on most naval aircraft perform many functions. Some systems operated by hydraulics are flight controls, landing gear, speed brakes, fixed-wing and rotary-wing folding mechanisms, auxiliary systems, and wheel brakes.

Hydraulics has many advantages as a power source for operating these units on aircraft.

- Hydraulics combine the advantages of lightweight, ease of installation, simplification of inspection, and minimum maintenance requirements.
- Hydraulics operation is almost 100-percent efficient, with only a negligible loss due to fluid friction.
- However, there are some disadvantages to using hydraulics.
- The possibility of leakage, both internal and external, may cause the complete system to become inoperative.
- Contamination by foreign matter in the system can cause malfunction of any unit. Cleanliness in hydraulics cannot be overemphasized.

## Components of a basic Hydraulic System

Basically, any hydraulic system contains the following units:

- A reservoir to hold a supply of hydraulic fluid.
- A pump to provide a flow of fluid.
- Tubing to transmit the fluid.
- A selector valve to direct the flow of fluid.
- An actuating unit to convert the fluid pressure into useful work.

A simple system using these essential units is shown in figure.

You can trace the flow of fluid from the reservoir through the pump to the selector valve. In figure, the flow of fluid created by the pump flows through the valve to the right end of the actuating cylinder. Fluid pressure forces the piston to the left. At the same time, the fluid that is on the left of the piston is forced out. It goes up through the selector valve and back to the reservoir through the return line.

When the selector valve is moved to the position indicated by the dotted lines, the fluid from the pump flows to the left side of the actuating cylinder. Movement of the piston can be stopped at any time simply by moving the selector valve to neutral. When the selector valve is in this position, all four ports are closed, and pressure is trapped in both working lines.

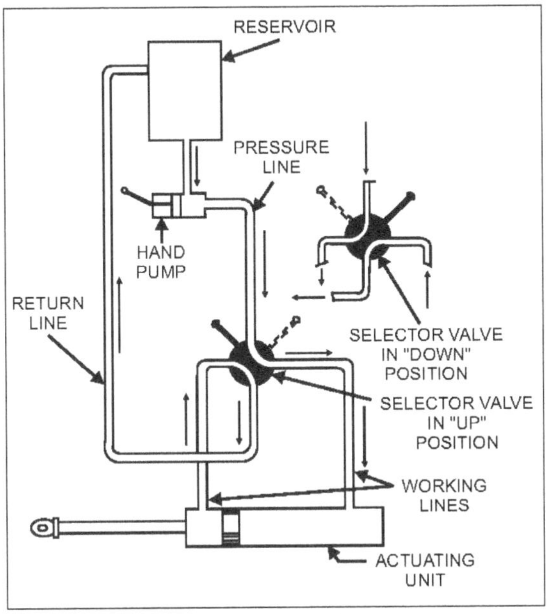


Basic hydraulic system, hand pump operated.

Figure shows a basic system with the addition of a power-driven pump and other essential components. These components are the filter, pressure regulator, accumulator, pressure gauge, relief valve, and two check valves. The function of these components is described below.

The filter removes foreign particles from the fluid, preventing moisture, dust, grit, and other undesirable matter from entering the system.

The pressure regulator unloads or relieves the power-driven pump when the desired pressure in the system is reached. Therefore, it is often referred to as an unloading valve. With none of the actuating units operating, the pressure in the line between the pump and selector valve builds up to the desired point. A valve in the pressure regulator automatically opens and fluid is bypassed back to the reservoir. (The bypass line is shown in figure, leading from the pressure regulator to the return line).

Many aircraft hydraulic systems do not use a pressure regulator. These systems use a pump that automatically adjusts to supply the proper volume of fluid as needed.

The accumulator serves a twofold purpose:

- It serves as a cushion or shock absorber by maintaining an even pressure in the system.
- It stores enough fluid under pressure to provide for emergency operation of certain actuating units.

The accumulator is designed with a compressed-air chamber separated from the fluid by a flexible diaphragm, or a removable piston.

The pressure gauge indicates the amount of pressure in the system.

The relief valve is a safety valve installed in the system. When fluid is bypassed through the valve to the return line, it returns to the reservoir. This action prevents excessive pressure in the system.

Check valves allow the flow of fluid in one direction only. There are numerous check valves installed at various points in the lines of all aircraft hydraulic systems. A careful study of figure shows why the two check valves are necessary in this system. One check valve prevents power pump pressure from entering the hand-pump line. The other valve prevents hand-pump pressure from being directed to the accumulator.

## Hydraulic Contamination

Hydraulic contamination is defined as foreign material in the hydraulic system of an aircraft. Foreign material might be grit, sand, dirt, dust, rust, water, or any other substance that is not soluble in the hydraulic fluid.

There are two basic ways to contaminate a hydraulic system. One is to inject particles, and the other is to intermix fluids, including water.

Particle contamination in a system may be self-generated through normal wear of system components. It is the injection of contaminants from outside that usually causes the most trouble. Regardless of its origin, any form of contamination in the hydraulic system will slow performance. In extreme cases, it seriously affects safety.

A single grain of sand or grit can cause internal failure of a hydraulic component. Usually, this type of contamination comes from poor servicing and fluid-handling procedures. For this reason, the highest level of cleanliness must be maintained when working on hydraulic components.

Only approved fill stand units are used to service naval aircraft hydraulic systems. By following a few basic rules, you can service hydraulic systems safely and keep contamination to a minimum.

- Never use fluid that has been left open for an undetermined period of time. Hydraulic fluid that is exposed to air will absorb dust and dirt.

- Never pour fluid from one container into another.
- Use only approved servicing units for the specific aircraft.
- Maintain hydraulic fluid-handling equipment in a high state of cleanliness.
- Always make sure you use the correct hydraulic fluid.

Contamination of the hydraulic system may be caused by wear or failure of hydraulic components and seals. This type of contamination is usually found through filter inspection and fluid analysis. Continued operation of a contaminated system may cause malfunctioning.

## Pneumatic Systems

There are two types of pneumatic systems currently used in naval aircraft. One type uses storage bottles for an air source, and the other has its own air compressor.

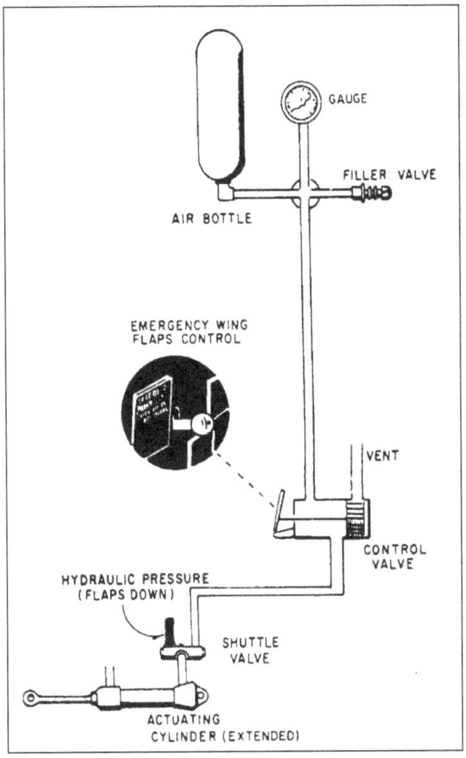


Emergency pneumatic system.

Generally, the storage bottle system is used only for emergency operation. This system has an air bottle, a control valve in the cockpit for releasing the contents of the cylinders, and a ground charge (filler) valve. The storage bottle must be filled with compressed air or nitrogen prior to flight. Air storage cylinder pneumatic systems are in use for emergency brakes, emergency landing gear extension, emergency flap extension, and for canopy release mechanisms.

When the control valve is properly positioned, the compressed air in the storage bottle is routed through the shuttle valve to the actuating cylinder.

The shuttle valve is a pressure-operated valve that separates the normal hydraulic system from the emergency pneumatic system. When the control handle is returned to the normal position, the air pressure in the lines is vented overboard through the vent port of the control valve.

The other type of pneumatic system in use has its own air compressor. It also has other equipment necessary to maintain an adequate supply of compressed air during flight. Most systems of this type must be serviced on the ground prior to flight. The air compressor used in most aircraft is driven by a hydraulic motor. Aircraft that have an air compressor use the compressed air for normal and emergency system operation.

# FUSELAGE

The fuselage is an aircraft's main body section. It holds crew, passengers, and cargo. In single-engine aircraft, it will usually contain an engine, as well, although in some amphibious aircraft the single engine is mounted on a pylon attached to the fuselage, which in turn is used as a floating hull. The fuselage also serves to position control and stabilization surfaces in specific relationships to lifting surfaces, which is required for aircraft stability and maneuverability.

## Materials

Inside view of the wooden, fabric-covered fuselage of a Fisher FP-202.

Early aircraft were constructed of wood frames covered in fabric. As monoplanes became popular, metal frames improved the strength, which eventually led to all-metal -structure aircraft, with metal covering for all its exterior surfaces - this was first pioneered in the second half of 1915. Some modern aircraft are constructed with composite materials for major control surfaces, wings, or the entire fuselage such as the Boeing 787. On the 787, it makes possible higher pressurization levels and larger windows for passenger comfort as well as lower weight to reduce operating costs. The Boeing 787 weighs 1500 lb less than if it were an all-aluminum assembly.

## Windows

Cockpit windshields on the Airbus A320 must withstand bird strikes up to 350 kt and are made of chemically strengthened glass. They are usually composed of three layers or plies, of glass or plastic: the inner two are 8 mm (0.3 in.) thick each and are structural, while the outer ply, about 3 mm thick, is a barrier against foreign object damage and abrasion, with often a hydrophobic coating. It must prevent fogging inside the cabin and de-ice from −50 °C (−58 °F), this was previously done with thin wires similar to a rear car window, but now is a transparent, nanometers thick coating of indium tin oxide siting between plies, electrically conductive and thus transmitting heat. Curved glass improves aerodynamics but sight criteria also needs larger panes. A cockpit windshield is composed of 4–6 panels, 35 kg (77 lb.) each on an Airbus A320. In its lifetime, an average aircraft goes through three or four windshields, and the market is shared evenly between OEM and higher margins aftermarket.

Cabin windows, made from much lighter than glass stretched acrylic glass, consists of multiple panes: An outer one built to support four times the maximum cabin pressure, an inner one for redundancy and a scratch pane near the passenger. Acrylic is susceptible to crazing: A network of fine cracks appears but can be polished to restore optical transparency, removal and polishing typically undergo every 2–3 years for uncoated windows.

## Wing Integration

"Flying wing" aircraft, such as the Northrop YB-49 Flying Wing and the Northrop B-2 Spirit bomber have no separate fuselage; instead what would be the fuselage is a thickened portion of the wing structure.

Conversely, there have been a small number of aircraft designs which have no separate wing, but use the fuselage to generate lift. Examples include National Aeronautics and Space Administration's experimental lifting body designs and the Vought XF5U-1 Flying Flapjack.

A blended wing body can be considered a mixture of the above. It carries the useful load in a fuselage producing lift. A modern example is Boeing X-48. One of the earliest aircraft using this design approach is Burnelli CBY-3, which fuselage was airfoil shaped to produce lift.

# WINGS

A swept wing KC-10 Extender (top) refuels a trapezoidal-wing F-22 Raptor.

A wing is a type of fin that produces lift, while moving through air or some other fluid. As such, wings have streamlined cross-sections that are subject to aerodynamic forces and act as airfoils. A wing's aerodynamic efficiency is expressed as its lift-to-drag ratio. The lift a wing generates at a given speed and angle of attack can be one to two orders of magnitude greater than the total drag on the wing. A high lift-to-drag ratio requires a significantly smaller thrust to propel the wings through the air at sufficient lift.

Lifting structures used in water, include various foils, such as hydrofoils. Hydrodynamics is the governing science, rather than aerodynamics. Applications of underwater foils occur in hydroplanes, sailboats and submarines.

## Aerodynamics

Condensation in the low pressure region over the wing of an Airbus A340, passing through humid air.

The design and analysis of the wings of aircraft is one of the principal applications of the science of aerodynamics, which is a branch of fluid mechanics. In principle, the properties of the airflow around any moving object can be found by solving the Navier-Stokes equations of fluid dynamics. However, except for simple geometries these equations are notoriously difficult to solve and simpler equations are used.

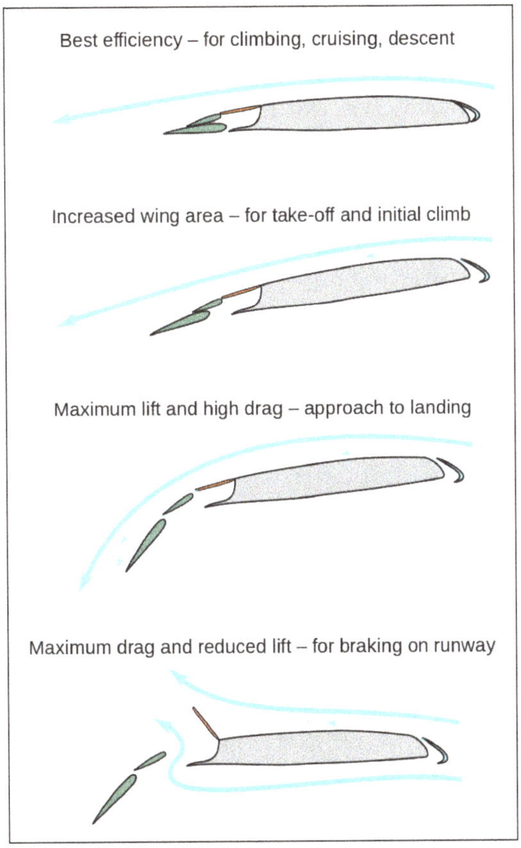


Flaps (green) are used in various configurations to increase the wing area and to increase the lift. In conjunction with spoilers (red), flaps maximize drag and minimize lift during the landing roll.

For a wing to produce *lift*, it must be oriented at a suitable angle of attack. When this occurs, the wing deflects the airflow downwards as it passes the wing. Since the wing exerts a force on the air to change its direction, the air must also exert an equal and opposite force on the wing, resulting in different air pressures over the surface of the wing. A region of lower-than-normal air pressure is generated over the top surface of the wing, with a higher pressure on the bottom of the wing. These air pressure differences can be measured directly using instrumentation or can be calculated from the airspeed distribution using basic physical principles such as Bernoulli's principle, which relates changes in air speed to changes in air pressure.

It is possible to calculate lift from: the pressure differences, the different velocities of the air above and below the wing, or from the total momentum change of the deflected air. Debates over which mathematical approach is the most convenient to use can be mistaken as differences of opinion about the basic principles of flight.

## Cross-sectional Shape

An airfoil is the shape of a wing, blade (of a propeller, rotor, or turbine), or sail. Wings with an asymmetrical cross section are the norm in subsonic flight. Wings with a symmetrical cross section can also generate lift by using a positive angle of attack to deflect air downward. Symmetrical airfoils have higher stalling speeds than cambered airfoils of the same wing area but are used in aerobatic aircraftas they provide practical performance whether the aircraft is upright or inverted. Another example comes from sailboats, where the sail is a thin membrane with no path-length difference between one side and the other.

For flight speeds near the speed of sound (transonic flight), airfoils with complex asymmetrical shapes are used to minimize the drastic increase in drag associated with airflow near the speed of sound. Such airfoils, called supercritical airfoils, are flat on top and curved on the bottom.

## Design Features

The wing of a landing BMI Airbus A319-100. The slats at its leading edge and the flaps at its trailing edge are extended.

Aircraft wings may feature some of the following:

- A rounded leading edge cross-section.
- A sharp trailing edge cross-section.
- Leading-edge devices such as slats, slots, or extensions.
- Trailing-edge devices such as flaps or flaperons (combination of flaps and ailerons).
- Winglets to keep wingtip vortices from increasing drag and decreasing lift.
- Dihedral, or a positive wing angle to the horizontal, increases *spiral stability* around the roll axis, whereas *anhedral*, or a negative wing angle to the horizontal, decreases spiral stability.

Aircraft wings may have various devices, such as flaps or slats that the pilot uses to modify the shape and surface area of the wing to change its operating characteristics in flight.

- Ailerons (usually near the wingtips) to roll the aircraft clockwise or counterclockwise about its long axis.
- Spoilers on the upper surface to disrupt the lift and to provide additional traction to an aircraft that has just landed but is still moving.
- Vortex generators to help prevent flow separation in transonic flow.
- Wing fences to keep flow attached to the wing by stopping boundary layer separation from spreading roll direction.
- Folding wings allow more aircraft storage in the confined space of the hangar deck of an aircraft carrier.
- Variable-sweep wing or "swing wings" that allow outstretched wings during low-speed flight (i.e., take-off and landing) and swept back wings for high-speed flight (including supersonic flight), such as in the F-111 Aardvark, the F-14 Tomcat, the Panavia Tornado, the MiG-23, the MiG-27, the Tu-160 and the B-1B Lancer warplanes.
- Strakes to improve flight characteristics.
- Chine, which may blend into the wing.
- Droop, a high-lift device.
- Fairings, structures whose primary function is to produce a smooth outline and reduce drag. For example, flap track fairings.

Wings may have other minor independent surfaces.

## Applications and Variants

Besides fixed-wing aircraft, applications for wing shapes include:

- Hang gliders, which use wings ranging from fully flexible (paragliders, gliding parachutes), flexible (framed sail wings), to rigid.
- Kites, which use a variety of lifting surfaces.
- Flying model airplanes.
- Helicopters, which use a rotating wing with a variable pitch angle to provide directional forces.
- Propellers, whose blades generate lift for propulsion.
- The NASA Space Shuttle, which uses its wings only to glide during its descent to a runway. These types of aircraft are called spaceplanes.

- Some racing cars, especially Formula One cars, which use upside-down wings (or *airfoils*) to provide greater traction at high speeds.
- Sailboats, which use sails as vertical wings with variable fullness and direction to move across water.

### In Nature

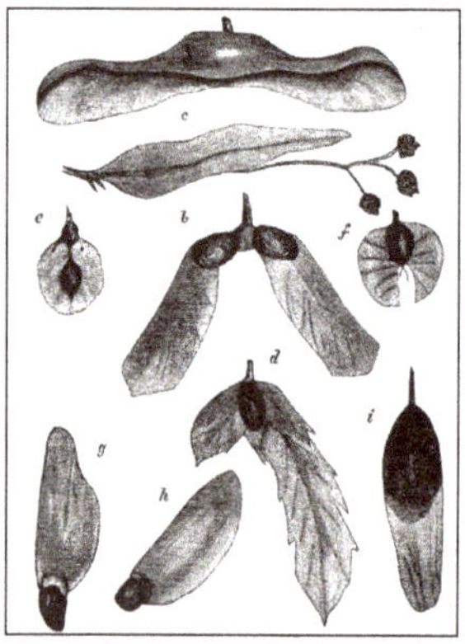

Winged tree seeds that cause autorotation in descent

A laughing gull, exhibiting the "gull wing" outline.

In nature, wings have evolved in insects, pterosaurs, dinosaurs (birds), and mammals (bats) as a means of locomotion. Various species of penguins and other flighted or flightless water birds such as auks, cormorants, guillemots, shearwaters, eider and scoter ducks and diving petrels are avid swimmers, and use their wings to propel through water.

Bat in flight.

## COCKPIT

A cockpit or flight deck is the area, usually near the front of an aircraft or spacecraft, from which a pilot controls the aircraft.

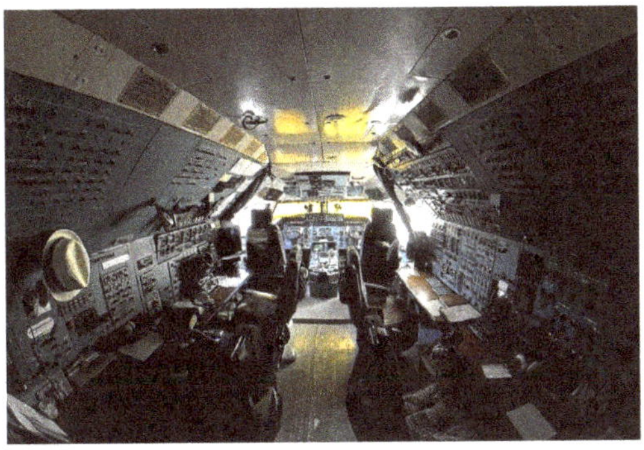

Cockpit of an Antonov An-124.

Cockpit of an A380. Most Airbus cockpits are glass cockpits featuring fly-by-wire technology.

The cockpit of an aircraft contains flight instruments on an instrument panel, and the controls that enable the pilot to fly the aircraft. In most airliners, a door separates the cockpit from the aircraft cabin. After the September 11, 2001 attacks, all major airlines fortified their cockpits against access by hijackers.

## Ergonomics

The first airplane with an enclosed cabin appeared in 1912 on the Avro Type F; however, during the early 1920s there were many passenger aircraft in which the crew remained open to the air while the passengers sat in a cabin. Military biplanes and the first single-engined fighters and attack aircraft also had open cockpits, some as late as the Second World War when enclosed cockpits became the norm.

The largest impediment to having closed cabins was the material used to make the windows. Prior to Perspex becoming available in 1933, windows were either safety glass, which was heavy, or cellulose nitrate (i.e. guncotton), which yellowed quickly and was extremely flammable. In the mid-1920s many aircraft manufacturers began using enclosed cockpits for the first time. Early airplanes with closed cockpits include the 1924 Fokker F.VII, the 1926 German Junkers W 34 transport, the 1926 Ford Trimotor, the 1927 Lockheed Vega, the Spirit of St. Louis and the passenger aircraft manufactured by the Douglas and Boeing companies during the mid-1930s. Open-cockpit airplanes were almost extinct by the mid-1950s, with the exception of training planes, crop-dusters and homebuilt aircraft designs.

Cockpit windows may be equipped with a sun shield. Most cockpits have windows that can be opened when the aircraft is on the ground. Nearly all glass windows in large aircraft have an anti-reflective coating, and an internal heating element to melt ice. Smaller aircraft may be equipped with a transparent aircraft canopy.

In most cockpits the pilot's control column or joystick is located centrally (centre stick), although in some military fast jets the side-stick is located on the right hand side. In some commercial airliners (i.e. Airbus—which features the glass cockpit concept) both pilots use a side-stick located on the outboard side, so Captain's side-stick on the left and First-officer's seat on the right.

Except for some helicopters, the *right seat* in the cockpit of an aircraft is the seat used by the co-pilot. The captain or pilot in command sits in the *left seat*, so that they can operate the throttles and other pedestal instruments with his right hand. The tradition has been maintained to this day, with the co-pilot on the right hand side.

The layout of the cockpit, especially in the military fast jet, has undergone standardisation, both within and between aircraft, manufacturers and even nations. An important development was the "Basic Six" pattern, later the "Basic T", developed from 1937 onwards by the Royal Air Force, designed to optimise pilot instrument scanning.

Ergonomics and Human Factors concerns are important in the design of modern cockpits. The layout and function of cockpit displays controls are designed to increase pilot situation awareness without causing information overload. In the past, many cockpits, especially in fighter aircraft, limited the size of the pilots that could fit into them. Now, cockpits are being designed to accommodate from the 1st percentile female physical size to the 99th percentile male size.

In the design of the cockpit in a military fast jet, the traditional "knobs and dials" associated with the cockpit are mainly absent. Instrument panels are now almost wholly replaced by electronic displays, which are themselves often re-configurable to save space. While some hard-wired dedicated switches must still be used for reasons of integrity and safety, many traditional controls are replaced by multi-function re-configurable controls or so-called "soft keys". Controls are incorporated onto the stick and throttle to enable the pilot to maintain a head-up and eyes-out position – the Hands On Throttle And Stick or HOTAS concept. These controls may be then further augmented by control media such as head pointing with a Helmet Mounted Sighting System or Direct voice input (DVI). Advances in auditory displays allow for Direct Voice Output of aircraft status information and for the spatial localisation of warning sounds for improved monitoring of aircraft systems.

The layout of control panels in modern airliners has become largely unified across the industry. The majority of the systems-related controls (such as electrical, fuel, hydraulics and pressurization) for example, are usually located in the ceiling on an overhead panel. Radios are generally placed on a panel between the pilot's seats known as the pedestal. Automatic flight controls such as the autopilot are usually placed just below the windscreen and above the main instrument panel on the glareshield. A central concept in the design of the cockpit is the Design Eye Position or "DEP", from which point all displays should be visible.

Most modern cockpits will also include some kind of integrated warning system.

In a 2013 comparative study of a number of novel methods for cockpit-user interaction, touchscreen produced the largest number of "best scores".

## Aerospace Industry Technologies

In the U.S. the Federal Aviation Administration (FAA) and the National Aeronautics and Space Administration (NASA) have researched the ergonomic aspects of cockpit design and have conducted investigations of airline industry accidents. Cockpit design disciplines include Cognitive science, Neuroscience, Human–computer interaction, Human Factors Engineering, Anthropometry and Ergonomics.

Aircraft designs have adopted the fully digital "glass cockpit". In such designs, instruments and gauges, including navigational map displays, use a user interface markup language known as ARINC 661. This standard defines the interface between an independent cockpit display system, generally produced by a single manufacturer, and the avionics equipment and user applications it is required to support, by means of displays and controls, often made by different manufacturers. The separation between the overall display system, and the applications driving it, allows for specialization and independence.

# EMPENNAGE

The empennage of a Boeing 747-200.

The empennage also known as the tail or tail assembly, is a structure at the rear of an aircraft that provides stability during flight, in a way similar to the feathers on an arrow. The term derives from the French language word *empenner* which means "to feather an arrow". Most aircraft feature an empennage incorporating vertical and horizontal stabilising surfaces which stabilise the flight dynamics of yaw and pitch, as well as housing control surfaces.

In spite of effective control surfaces, many early aircraft that lacked a stabilising empennage were virtually unflyable. Even so-called "tailless aircraft" usually have a tail fin (usually a vertical stabiliser). Heavier-than-air aircraft without any kind of empennage (such as the Northrop B-2) are rare.

## Structure

Structurally, the empennage consists of the entire tail assembly, including the tailfin, the tailplane and the part of the fuselage to which these are attached. On an airliner this would be all the flying and control surfaces behind the rear pressure bulkhead.

The front (usually fixed) section of the tailplane is called the tailplane or horizontal stabiliser and is used to provide pitch stability. The rear section is called the elevator, and is usually hinged to the horizontal stabiliser. The elevator is a movable aerofoil that controls changes in pitch, the up-and-down motion of the aircraft's nose. Some aircraft employ an all-moving stabiliser and elevators in one unit, known as a stabilator or "full-flying stabiliser".

The vertical tail structure (or fin) has a fixed front section called the vertical stabiliser, used to restrict side-to-side motion of the aircraft (yawing). The rear section of the vertical fin is the rudders, a movable aerofoil that is used to turn the aircraft's nose to one side or the other. When used in combination with the ailerons, the result is a banking turn, often referred to as a "coordinated turn".

Some aircraft are fitted with a tail assembly that is hinged to pivot in two axes forward of the fin and stabiliser, in an arrangement referred to as a *movable tail*. The entire empennage is rotated vertically to actuate the horizontal stabiliser, and sideways to actuate the fin.

The aircraft's cockpit voice recorder, flight data recorder and emergency locator transmitter (ELT) are often located in the empennage, because the aft of the aircraft provides better protection for these in most aircraft crashes.

## Trim

In some aircraft trim devices are provided to eliminate the need for the pilot to maintain constant pressure on the elevator or rudder controls.

The trim device may be:

- A trim tab on the rear of the elevators or rudder which act to change the aerodynamic load on the surface. Usually controlled by a cockpit wheel or crank.
- An adjustable stabiliser into which the stabiliser may be hinged at its spar and adjustably jacked a few degrees in incidence either up or down. Usually controlled by a cockpit crank.
- A bungee trim system which uses a spring to provide an adjustable preload in the controls. Usually controlled by a cockpit lever.
- An anti-servo tab used to trim some elevators and stabilators as well as increased control force feel. Usually controlled by a cockpit wheel or crank.

- A servo tab used to move the main control surface, as well as act as a trim tab. Usually controlled by a cockpit wheel or crank.

Multi-engined aircraft often have trim tabs on the rudder to reduce the pilot effort required to keep the aircraft straight in situations of asymmetrical thrust, such as single engine operations.

## Tail Configurations

Aircraft empennage designs may be classified broadly according to the fin and tailplane configurations.

The overall shapes of individual tail surfaces (tailplane planforms, fin profiles) are similar to wing planforms.

## Tailplanes

The tailplane comprises the tail-mounted fixed horizontal stabiliser and movable elevator. Besides its planform, it is characterised by:

- Number of tailplanes: From 0 (tailless or canard) to 3 (Roe triplane).
- Location of tailplane: Mounted high, mid or low on the fuselage, fin or tail booms.
- Fixed stabiliser and movable elevator surfaces, or a single combined stabilator or *flying tail* (General Dynamics F-111 Aardvark).

Some locations have been given special names:

- Cruciform tail: The horizontal stabilisers are placed midway up the vertical stabiliser, giving the appearance of a cross when viewed from the front. Cruciform tails are often used to keep the horizontal stabilisers out of the engine wake, while avoiding many of the disadvantages of a T-tail. Examples include the Hawker Sea Hawk and Douglas A-4 Skyhawk.
- T-tail: The horizontal stabiliser is mounted on top of the fin, creating a "T" shape when viewed from the front. T-tails keep the stabilisers out of the engine wake, and give better pitch control. T-tails have a good glide ratio, and are more efficient on low speed aircraft. However, the T-tail has several disadvantages. It is more likely to enter a deep stall, and is more difficult to recover from a spin. For this reason a small secondary stabiliser or tail-let may be fitted lower down where it will be in free air when the aircraft is stalled. A T-tail must be stronger, and therefore heavier than a conventional tail. T-tails also tend to have a larger radar cross section. Examples include the Gloster Javelin and McDonnell Douglas DC-9.

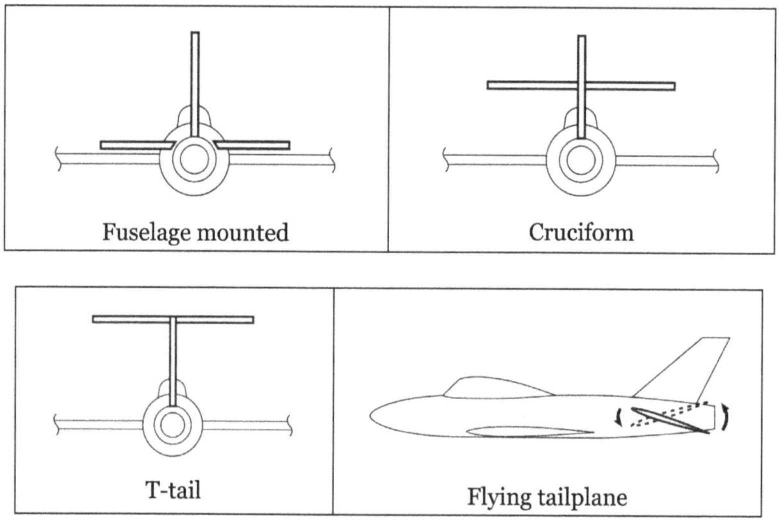

Fuselage mounted | Cruciform | T-tail | Flying tailplane

## Fins

The fin comprises the fixed vertical stabiliser and rudder. Besides its profile, it is characterised by:

- Number of fins: Usually one or two.
- Location of fins: On the fuselage (over or under), tailplane, tail booms or wings.

Twin fins may be mounted at various points:

- Twin tail: A twin tail, also called an H-tail, consists of two small vertical stabilisers on either side of the horizontal stabiliser. Examples include the Antonov An-225 Mriya, B-25 Mitchell, Avro Lancaster, and ERCO Ercoupe.
- Twin boom: A twin boom has two fuselages or booms, with a vertical stabiliser on each, and a horizontal stabiliser between them. Examples include the P-38 Lightning, de Havilland Vampire, Sadler Vampire, and Edgley Optica.
- Wing mounted midwing as on the F7U Cutlass or on the wing tips as on the Handley Page Manx and Rutan Long-EZ.

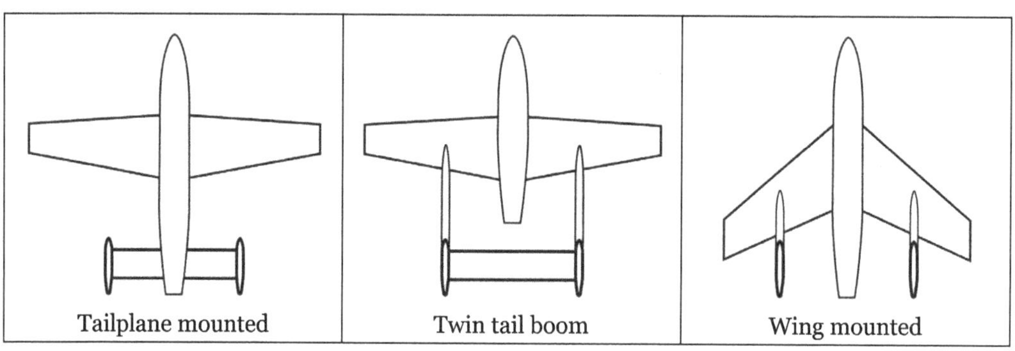

Tailplane mounted | Twin tail boom | Wing mounted

Unusual fin configurations include:

- No fin: As on the McDonnell Douglas X-36. This configuration is sometimes incorrectly referred to as "tailless".
- Multiple fins: Examples include the Lockheed Constellation (three), Bellanca 14-13 (three), and the Northrop Grumman E-2 Hawkeye (four).
- Ventral fin: Underneath the fuselage. Often used in addition to a conventional fin as on the (North American X-15 and Dornier Do 335).

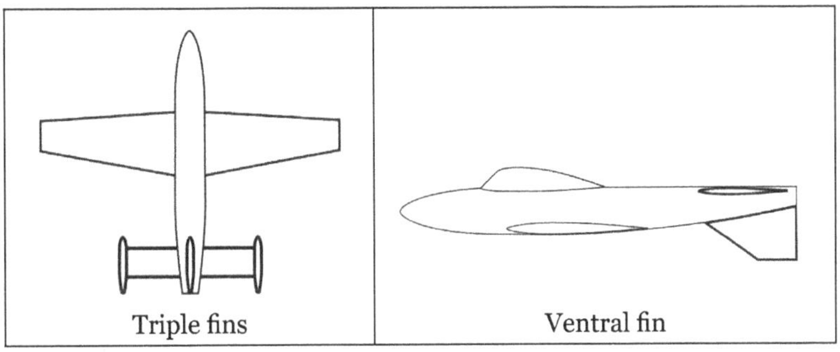


## V and X Tails

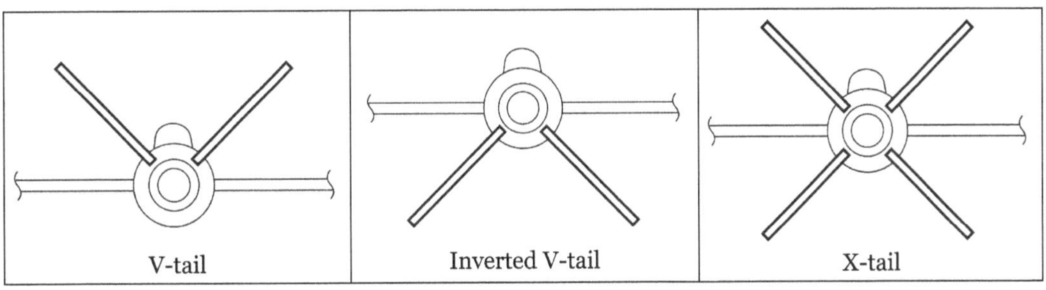


An alternative to the fin-and-tailplane approach is provided by the V-tail and X-tail designs. Here, the tail surfaces are set at diagonal angles, with each surface contributing to both pitch and yaw. The control surfaces, sometimes called ruddervators, act differentially to provide yaw control (in place of the rudder) and act together to provide pitch control (in place of the elevator).

- V tail: A V-tail can be lighter than a conventional tail in some situations and produce less drag, as on the Fouga Magister trainer, Northrop Grumman RQ-4 Global Hawk RPV and X-37 spacecraft. A V-tail may also have a smaller radar signature. Other aircraft featuring a V-tail include the Beechcraft Model 35 Bonanza, and Davis DA-2. A slight modification to the V-tail can be found on the Waiex and Monnett Moni called a Y-tail.
- Inverted V tail:The unmanned Predator uses an inverted V-tail as do the Lazair and Mini-IMP.

- X tail: The Lockheed XFV and Convair XFY Pogo both featured "X" tails, which were reinforced and fitted with a wheel on each surface so that the craft could sit on its tail and take off and land vertically.

## Outboard Tail

SpaceShipOne at the US National Air and Space Museum.

An outboard tail is split in two, with each half mounted on a short boom just behind and outboard of each wing tip. It comprises outboard horizontal stabilizers (OHS) and may or may not include additional boom-mounted vertical stabilizers (fins). In this position, the tail surfaces interact constructively with the wingtip vortices and, with careful design, can significantly reduce drag to improve efficiency, without adding unduly to the structural loads on the wing.

The configuration was first developed during World War II by Richard Vogt and George Haag at Blohm & Voss. The Skoda-Kauba SL6 tested the proposed control system in 1944 and, following several design proposals, an order was received for the Blohm & Voss P 215 just weeks before the war ended. The outboard tail reappeared on the Scaled Composites SpaceShipOne in 2003 and SpaceShipTwo in 2010.

## Tailless Aircraft

The DH108 *Swallow*.

A tailless aircraft (often tail-less) traditionally has all its horizontal control surfaces on its main wing surface. It has no horizontal stabiliser - either tailplane or canard foreplane (nor does it have a second wing in tandem arrangement). A 'tailless' type usually still has a vertical stabilising fin (vertical stabiliser) and control surface (rudder). However, NASA adopted the 'tailless' description for the novel X-36 research aircraft which has a canard foreplane but no vertical fin.

The most successful tailless configuration has been the tailless delta, especially for combat aircraft.

# TAILPLANE

The tailplane of this P-51 is shown in pink.

A tailplane, also known as a horizontal stabiliser, is a small lifting surface located on the tail (empennage) behind the main lifting surfaces of a fixed-wing aircraft as well as other non-fixed-wing aircraft such as helicopters and gyroplanes. Not all fixed-wing aircraft have tailplanes. Canards, tailless and flying wing aircraft have no separate tailplane, while in V-tail aircraft the vertical stabiliser, rudder, and the tail-plane and elevator are combined to form two diagonal surfaces in a V layout.

The function of the tailplane is to provide stability and control. In particular, the tailplane helps adjust for changes in position of the centre of pressure or centre of gravity caused by changes in speed and attitude, fuel consumption, or dropping cargo or payload.

## Tailplane Types

The tailplane comprises the tail-mounted fixed horizontal stabiliser and movable elevator. Besides its planform, it is characterised by:

- Number of tailplanes: From 0 (tailless or canard) to 3 (Roe triplane).
- Location of tailplane: Mounted high, mid or low on the fuselage, fin or tail booms.
- Fixed stabiliser and movable elevator surfaces, or a single combined stabilator or *(all) flying tail.*

Some locations have been given special names:

- Cruciform: Mid-mounted on the fin.
- T-tail: High-mounted on the fin.

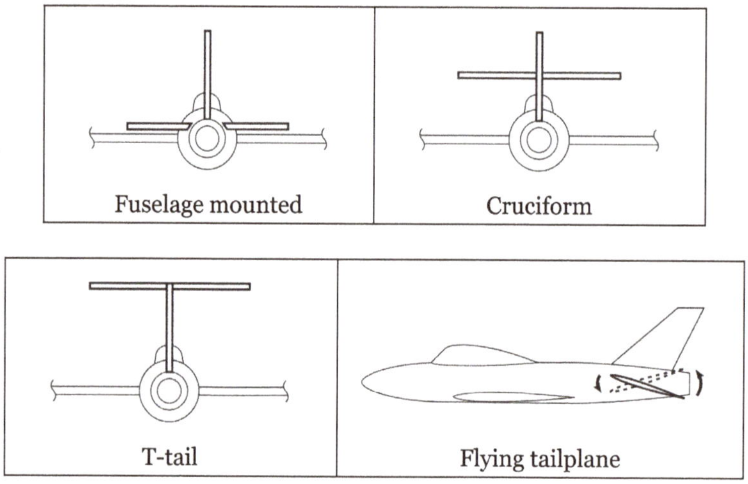


## Stability

Tailplane (in shadow) of an easyJet Airbus A319.

A wing with a conventional aerofoil profile makes a negative contribution to longitudinal stability. This means that any disturbance (such as a gust) which raises the nose produces a nose-up pitching moment which tends to raise the nose further. With the same disturbance, the presence of a tailplane produces a restoring nose-down pitching moment, which may counteract the natural instability of the wing and make the aircraft longitudinally stable (in much the same way a weather vane always points into the wind).

The longitudinal stability of an aircraft may change when it is flown "hands-off"; i.e. when the flight controls are subject to aerodynamic forces but not pilot input forces.

## Damping

In addition to giving a restoring force (which on its own would cause oscillatory motion) a tailplane gives damping. This is caused by the relative wind seen by the tail as the aircraft rotates around the centre of gravity. For example, when the aircraft is oscillating, but is momentarily aligned with the overall vehicle's motion, the tailplane still sees a relative wind that is opposing the oscillation.

## Lift

Depending on the aircraft design and flight regime, its tailplane may create positive lift or negative lift (downforce). It is sometimes assumed that on a stable aircraft this will always be a net down force, but this is untrue.

On some pioneer designs, such as the Bleriot XI, the centre of gravity was between the neutral point and the tailplane, which also provided positive lift. However this arrangement can be unstable and these designs often had severe handling issues. The requirements for stability were not understood until shortly before World War I - the era within which the British Bristol Scout light biplane was designed for civilian use, with an airfoiled lifting tail throughout its production run into the early World War I years and British military service from 1914-1916 — when it was realised that moving the centre of gravity further forwards allowed the use of a non-lifting tailplane in which the lift is nominally neither positive nor negative but zero, which leads to more stable behaviour. Later examples of aircraft from World War I and onwards into the interwar years that had positive lift tailplanes include, chronologically, the Sopwith Camel, Charles Lindbergh's Spirit of St. Louis, the Gee Bee Model R Racer - all aircraft with a reputation for being difficult to fly, and the easier-to-fly Fleet Finch two-seat Canadian trainer biplane, itself possessing a flat-bottom airfoiled tailplane unit not unlike the earlier Bristol Scout. But with care a lifting tailplane can be made stable. An example is provided by the Bachem Ba 349 Natter VTOL rocket-powered interceptor, which had a lifting tail and was both stable and controllable in flight.

In many modern conventional aircraft, the centre of gravity is placed ahead of the neutral point.The wing lift then exerts a pitch-down moment around the centre of gravity, which must be balanced by a pitch-up moment (implying negative lift) from the tailplane. A disadvantage is that it generates trim drag.

## Active Stability

Using a computer to control the elevator allows aerodynamically unstable aircraft to be flown in the same manner.

Aircraft such as the F-16 are flown with artificial stability. The advantage of this is a significant reduction in drag caused by the tailplane, and improved maneuverability.

### Mach Tuck

At transonic speeds, an aircraft can experience a shift rearwards in the center of pressure due to the buildup and movement of shockwaves. This causes a nose-down pitching moment called Mach tuck. Significant trim force may be needed to maintain equilibrium, and this is most often provided using the whole tailplane in the form of an all-flying tailplane or stabilator.

### Control

A tailplane usually has some means allowing the pilot to control the amount of lift produced by the tailplane. This in turn causes a nose-up or nose-down pitching moment on the aircraft, which is used to control the aircraft in pitch. Elevator A conventional tailplane normally has a hinged aft surface called an elevator.

Stabilator or all-moving tail In transonic flight shock waves generated by the front of the tailplane render any elevator unusable. An all-moving tail was developed by the British for the Miles M.52, but first saw actual transonic flight on the Bell X-1; Bell Aircraft Corporation had included an elevator trim device that could alter the angle of attack of the entire tailplane. This saved the program from a costly and time-consuming rebuild of the aircraft.

Transonic and supersonic aircraft now have all-moving tailplanes to counteract Mach tuck and maintain maneuverability when flying faster than the critical Mach number. Normally called a stabilator, this configuration is often referred to as an "all-moving" or "all-flying" tailplane.

## LANDING GEAR

Landing gear is the undercarriage of an aircraft or spacecraft and may be used for either takeoff or landing. For aircraft it is generally needed for both. It was also formerly called alighting gear by some manufacturers, such as the Glenn L. Martin Company.

For aircraft, the landing gear supports the craft when it is not flying, allowing it to take off, land, and taxi without damage. Wheeled landing gear is the most common with skis or floats needed to operate from snow/ice/water and skids for vertical operation on land. Faster aircraft have retractable undercarriages, which fold away during flight to reduce drag.

Some unusual landing gear have been evaluated experimentally. These include: no landing gear (to save weight), made possible by operating from a catapult cradle and flexible landing deck: air cushion (to enable operation over a wide range of ground obstacles and water/snow/ice); tracked (to reduce runway loading).

For launch vehicles and spacecraft landers, the landing gear usually only supports the vehicle on landing, and is not used for takeoff or surface movement.

## Aircraft

Oleo (pressurized gas - oil) strut with torque links.

Oleo strut with trailing link.

Methods of landing gear shock absorption: Aircraft landing gear includes wheels equipped with solid shock absorbers on light planes, and air/oil oleo struts on larger aircraft. Skis are used for operating from snow and floats from water. (Helicopters use skids, pontoons or wheels depending on their size and role).

Rubber discs in compression with trailing link, used on Mooney M20 series of aircraft.

Simple steel tube.

The landing gear represents 2.5 to 5% of the MTOW and 1.5 to 1.75% of the aircraft cost but 20% of the airframe direct maintenance cost. A suitably-designed wheel can support 30 t (66,000 lb), tolerate a ground speed of 300 km/h and roll a distance of 500,000 km (310,000 mi); it has a 20,000 hours time between overhaul and a 60,000 hours or 20 years life time.

## Gear Arrangements

Wheeled undercarriages normally come in two types:

- Conventional or "taildragger" undercarriage, where there are two main wheels towards the front of the aircraft and a single, much smaller, wheel or skid at the rear.
- Tricycle undercarriage, where there are two main wheels (or wheel assemblies) under the wings and a third smaller wheel in the nose.

The taildragger arrangement was common during the early propeller era, as it allows more room for propeller clearance. Most modern aircraft have tricycle undercarriages. Taildraggers are considered harder to land and take off (because the arrangement is usually *unstable*, that is, a small deviation from straight-line travel will tend to increase rather than correct itself), and usually require special pilot training. A small tail wheel or skid/bumper may be added to a tricycle undercarriage to prevent damage to the underside of the fuselage if over-rotation occurs on take-off leading to a tail strike. Aircraft with tail-strike protection include the B-29 Superfortress, Boeing 727 trijet and Concorde. Some aircraft with retractable conventional landing gear have a fixed tailwheel. Hoerner estimated the drag of the Me-109 fixed tailwheel and compared it with that of other protrusions such as the pilot's canopy.

A third arrangement (known as tandem or bicycle) has the main and nose gear located fore and aft of the center of gravity under the fuselage with outriggers on the wings. This is used when there is no convenient location on either side of the fuselage to attach the main undercarriage or to store it when retracted. Examples include the Lockheed

U-2 spy plane and the Harrier Jump Jet. The B-52 bomber uses a similar arrangement, except that the fore and aft gears each have two twin-wheel units side by side.

Conventional undercarriage (SAN Jodel D.140 Mousquetaire).

Tricycle undercarriage (Mooney M20J).

Quadricycle gear is similar to bicycle but with two sets of wheels displaced laterally in the fore and aft positions. Raymer classifies the B-52 gear as quadricycle. The experimental Fairchild XC-120 Packplane had quadricycle gear located in the engine nacelles to allow unrestricted access beneath the fuselage for attaching a large freight container.

## Retractable Gear

To decrease drag in flight undercarriages retract into the wings and/or fuselage with wheels flush with the surrounding surface or concealed behind flush-mounted doors; this is called *retractable gear*. If the wheels don't retract completely but protrude partially exposed to the airstream, it is called a semi-retractable gear.

Most retractable gear is hydraulically operated, though some is electrically operated or even manually operated on very light aircraft. The landing gear is stowed in a compartment called a wheel well.

The landing gear of a Boeing 767 retracting into the fuselage.

Pilots confirming that their landing gear is down and locked refer to "three greens" or "three in the green.", a reference to the electrical indicator lights (or painted panels of mechanical indicator units) from the nosewheel/tailwheel and the two main gears. Blinking green lights or red lights indicate the gear is in transit and neither up and locked or down and locked. When the gear is fully stowed up with the up-locks secure, the lights often extinguish to follow the dark cockpit philosophy; some airplanes have gear up indicator lights.

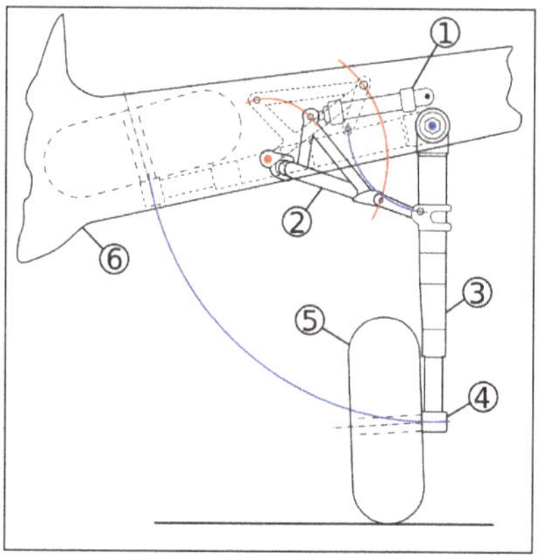


Schematic showing hydraulically operated landing gear, with the wheel stowed in the wing root of the aircraft.

Redundant systems are used to operate the landing gear and redundant main gear legs may also be provided so the aircraft can be landed in a satisfactory manner in a range of failure scenarios. The Boeing 747 was given four separate and independent hydraulic systems (when previous airliners had two) and four main landing gear posts (when previous airliners had two). Safe landing would be possible if two main gear legs were torn off provided they were on opposite sides of the fuselage. In the case of power failure in a light aircraft, an emergency extension system is always available. This may be a manually operated crank or pump, or a mechanical free-fall mechanism which disengages the uplocks and allows the landing gear to fall under gravity.

A Boeing 737-700 with main undercarriage retracted in the wheel wells without landing gear doors.

## Large Aircraft

As aircraft weights have increased more wheels have been added and runway thickness has increased to keep within the runway loading limit. This progression is shown by Torenbeek.

A Lockheed JetStar, with 4 wheels supporting about 40,000lb, needed a 10in thick flexible (asphalt) pavement. The thickness rose to 25in for a McDonnell Douglas DC-10/10 with about 400,000lb supported on 8 wheels. Heavier aircraft were able to operate from the same thickness pavements by using more wheels. For example the heavier McDonnell Douglas DC-10/30 had 10 wheels and the first Boeing 747, weighing about 700,000lb, had 16 wheels. The same-weight Lockheed C-5, with 24 wheels, only needs an 18in pavement.

The earliest "giant" aircraft placed in quantity production, the Zeppelin-Staaken R.VI German World War I long-range bomber of 1916, used eighteen wheels for its undercarriage, split between two wheels on its nose gear struts, and sixteen wheels on its main gear units — split into four side-by-side quartets each, two quartets of wheels per side — under each tandem engine nacelle, to support its loaded weight of almost 12 metric tons. Multiple "tandem wheels" on an aircraft — particularly for cargo aircraft, mounted to the fuselage lower sides as retractable main gear units on modern designs — were first seen during World War II, on the experimental German Arado Ar 232 cargo aircraft, which used a row of eleven "twinned" fixed wheel sets directly under the fuselage centerline to handle heavier loads while on the ground. Many of today's large cargo aircraft use this arrangement for their retractable main gear setups (usually mounted on the lower corners of the central fuselage structure). The Airbus A340-500/-600 has an additional four-wheel undercarriage bogie on the fuselage centerline, much like the twin-wheel unit in the same general location, used on later DC-10 and MD-11 airliners.

The prototype Convair XB-36, a very heavy aircraft for its day, had most of its weight on two main wheels which needed runways at least 22 inches thick. There were only three in the world. Production aircraft used two 4-wheel bogies which allowed the aircraft to use any airfield suitable for a B-29.

The Boeing 747 has five sets of wheels: a nosewheel assembly and four sets of four-wheel bogies. A set is located under each wing, and two inner sets are located in the fuselage, a little rearward of the outer bogies, adding up to a total of eighteen wheels and tires. The Airbus A380 also has a four-wheel bogie under each wing with two sets of six-wheel bogies under the fuselage.

The world's largest jet cargo aircraft, the Soviet Antonov An-225 has 4 wheels on the twin-strut nose gear units (as its smaller "stablemate", the Antonov An-124 also uses), and 28 main gear wheel/tire units, adding up to a total of 32 wheels and tires.

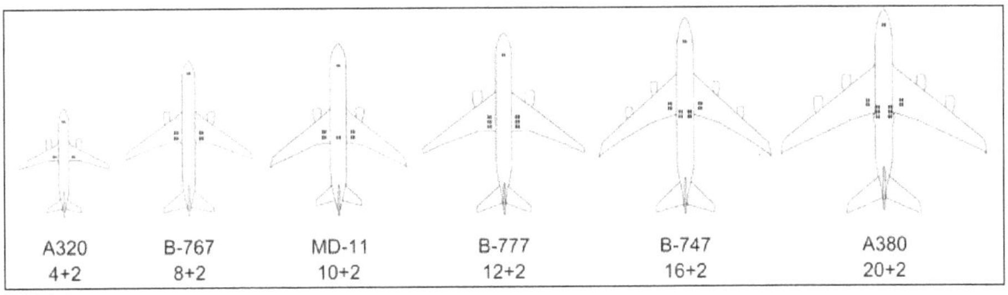


Wheel arrangements of large airliners.

Oleo strut rear landing gear of an Antonov An-124 (24 wheel landing gear).

The A340-600 has an additional main undercarriage on the fuselage belly.

## STOL Aircraft

STOL aircraft have a higher sink-rate requirement if a carrier-type, no-flare landing technique has to be adopted to reduce touchdown scatter. For example the Saab 37 Viggen, with landing gear designed for a 5m/sec impact, could use a carrier-type landing and HUD to reduce its scatter from 300 m to 100m.

The de Havilland Canada DHC-4 Caribou used long-stroke legs to land from a steep approach with no float.

## Operation from Water

The weight of a landplane can be accommodated with suitable pavement construction and landing gear design. Aircraft operation from water is limited by sea state, or wave height. With conventional hulls light seaplanes are limited to lower wave heights than heavier ones. Stinton shows an approximate relationship where a 2,000 lb aircraft is limited to wave heights of about 1 ft rising to a 100,000 lb aircraft wave height of about 6 ft. An exception is the ShinMaywa US-2 flying boat, with a sea-operating weight of about 80,000 lb, which has made successful landings and take-offs with wave heights of 13 feet.

A flying boat has a lower fuselage with the shape of a boat hull giving it buoyancy, usually with a "step" near the center of gravity to allow the aircraft to more easily break free of the water's surface for takeoff. Wing-mounted floats, or more rarely, stub-wing like sponsons are added for stability. Sponsons are attached to the lower sides of the fuselage, with their lower surfaces even with the chines forming the longitudinal lower corners of a flying boat's lower hull contours.

A floatplane has landing gear comprising two or more streamlined floats. Amphibious floats have retractable wheels for land operation.

An amphibious aircraft or amphibian usually has two distinct landing gears, namely a "boat" hull/floats and retractable wheels, which allow it to operate from land or water. In the 1970's trials were done with a truly amphibious landing gear, an air cushion, fitted to a Lake LA-4 and a de Havilland Canada DHC-5 Buffalo.

The Convair F2Y Sea Dart prototype seaplane fighter used retractable skis

Beaching gear is detachable wheeled landing gear that allows a non-amphibious floatplane or flying boat to be maneuvered on land. It is typically used only for aircraft maintenance and storage and is carried in the aircraft's cargo compartment or stored elsewhere when not in use. Beaching gear may consist of individual detachable wheels or a cradle that supports the entire aircraft. In the former case, the beaching gear is manually attached or detached with the aircraft in the water; in the latter case, the aircraft is maneuvered onto the cradle. Helicopters able to land on water may have floats or a hull.

Cessna 208 floatplane with amphibious floats.

Amphibious Consolidated PBY Catalina with landing gear lowered.

Non-amphibious Short Solent flying boat with beaching gear.

## Shipboard Operation

The landing gear on fixed-wing aircraft that land on aircraft carriers have a higher sink-rate requirement because the aircraft are flown onto the deck with no landing flare.

Other features have related to catapult take-off requirements for specific aircraft. For example, the Buccaneer was pulled down onto its tail-skid to set the required nose-up attitude. The naval McDonnell Douglas F-4 Phantom II in UK service needed an extending nosewheel leg to set the wing attitude at launch.

The landing gear for an aircraft using a (ski-jump (aviation)) on take-off is subjected to loads of 0.5g which also last for much longer than a landing impact.

Helicopters may have a deck-lock harpoon to anchor them to the deck.

## In-flight Use

Some aircraft have a requirement to use the landing-gear as a speed brake.

Flexible mounting of the stowed main landing-gear bogies on the Tupolev Tu-22R raised the aircraft flutter speed to 550 kts. The bogies oscillated within the nacelle under the control of dampers and springs as an anti-flutter device.

## Gear Common to Different Aircraft

Some experimental aircraft have used gear from existing aircraft to reduce program costs. The Martin-Marietta X-24 lifting body used the nose/main gear from the North American T-39 / Northrop T-38 and the Grumman X-29 from the Northrop F-5 / General Dynamics F-16.

## Other Types

### Skis

Wheel-skis.

When an airplane needs to land on surfaces covered by snow, the landing gear usually consists of skis or a combination of wheels and skis.

## Detachable

Some aircraft use wheels for takeoff and jettison them when airborne for improved streamlining without the complexity, weight and space requirements of a retraction mechanism. The wheels are sometimes mounted onto axles that are part of a separate "dolly" (for main wheels only) or "trolley" (for a three-wheel set with a nosewheel) chassis. Landing is done on skids or similar simple devices.

Historical examples include the "dolly"-using Messerschmitt Me 163 *Komet* rocket fighter, the Messerschmitt Me 321 *Gigant* troop glider, and the first eight "trolley"-using prototypes of the Arado Ar 234 jet reconnaissance bomber. The main disadvantage to using the takeoff dolly/trolley and landing skid(s) system on German World War II aircraft – intended for a sizable number of late-war German jet and rocket-powered military aircraft designs – was that aircraft would likely be scattered all over a military airfield after they had landed from a mission, and would be unable to taxi on their own to an appropriately hidden "dispersal" location, which could easily leave them vulnerable to being shot up by attacking Allied fighters. A related contemporary example are the wingtip support wheels ("pogos") on the Lockheed U-2 reconnaissance aircraft, which fall away after take-off and drop to earth; the aircraft then relies on titanium skids on the wingtips for landing.

## Rearwards and Sideways Retraction

A Royal Air Force P-47 with its raked-forward main gear, and rearward-angled main wheel position (when retracted) indicated by the just-visible open wheel door.

Some main landing gear struts on World War II aircraft, in order to allow a single-leg main gear to more efficiently store the wheel within either the wing or an engine nacelle, rotated the single gear strut through a 90° angle during the rearwards-retraction sequence to allow the main wheel to rest "flat" above the lower end of the main gear strut, or flush within the wing or engine nacelles, when fully retracted. Examples are the Curtiss P-40, Vought F4U Corsair, Grumman F6F Hellcat, Messerschmitt Me 210 and Junkers Ju 88. The Aero Commander family of twin-engined business aircraft also shares this feature on the main gears, which retract aft into the ends of the engine nacelles. The rearward-retracting nosewheel strut on the Heinkel He 219 and the forward-retracting nose gear strut on the later Cessna Skymaster similarly rotated 90 degrees as they retracted.

On most World War II single-engined fighter aircraft (and even one German heavy bomber design) with sideways retracting main gear, the main gear that retracted into the wings was meant to be raked forward, towards the aircraft's nose in the "down" position for better ground handling, with a retracted position that placed the main wheels at some angle "behind" the main gear's attachment point to the airframe – this led to a complex angular geometry for setting up the "pintle" angles at the top ends of the struts for the retraction mechanism's axis of rotation, with some aircraft, like the P-47 Thunderbolt and Grumman Bearcat, even mandating that the main gear struts lengthen as they were extended down from the wings to assure proper ground clearance for their large four-bladed propellers. One exception to the need for this complexity in many WW II fighter aircraft was Japan's famous Zero fighter, whose main gear stayed at a perpendicular angle to the centerline of the aircraft when extended, as seen from the side.

## Variable Axial Position of Main Wheels

The main wheels on the Vought F7U Cutlass could move 20 inches between a forward and aft position. The forward position was used for take-off to give a longer lever-arm for pitch control and greater nose-up attitude. The aft position was used to reduce landing bounce and reduce risk of tip-back during ground handling.

## Tandem Layout

Hawker Siddeley Harrier GR7 (ZG472). The two mainwheels are in line astern under the fuselage, with a smaller wheel on each wing.

The *tandem* or bicycle layout is used on the Hawker Siddeley Harrier, which has two mainwheels in line astern under the fuselage and a smaller wheel near the tip of each wing. On second generation Harriers, the wing is extended past the outrigger wheels to allow greater wing-mounted munition loads to be carried, or to permit wing-tip extensions to be bolted on for ferry flights.

A tandem layout was evaluated by Martin using a Martin A-26 to support its use on Martin's first jet bomber, the Martin XB-48. This configuration proved so manoeuvrable that it was also selected for the B-47 Stratojet. It was also used on the U-2, Myasishchev M-4, Yakovlev Yak-25, Yak-28, Sud Aviation Vautour. A variation of the multi tandem layout is also used on the B-52 Stratofortress which has four main wheel bogies (two forward and two aft) underneath the fuselage and a small outrigger wheel supporting each wing-tip. The B-52's landing gear is also unique in that all four pairs of main wheels can be steered. This allows the landing gear to line up with the runway and thus makes crosswind landings easier (using a technique called *crab landing*). Since tandem aircraft cannot rotate for takeoff, the forward gear must be long enough to give the wings the correct angle of attack during takeoff. During landing, the forward gear must not touch the runway first, otherwise the rear gear will slam down and may cause the aircraft to bounce and become airborne again.

## Crosswind Landing Accommodation

The "castoring" main gear arrangement on a Blériot XI.

One very early undercarriage incorporating castoring for crosswind landings was pioneered on the Bleriot VIII design of 1908. It was later used in the much more famous Blériot XI Channel-crossing aircraft of 1909 and also copied in the earliest examples of the Etrich Taube. In this arrangement the main landing gear's shock absorption was taken up by a vertically sliding bungee cord-sprung upper member. The vertical post along which the upper member slid to take landing shocks also had its lower end as the rotation point for the forward end of the main wheel's suspension fork, allowing the main gear to pivot on moderate crosswind landings.

Manually-adjusted main-gear units on the B-52 can be set for crosswind take-offs. It rarely has to be used from SAC-designated airfields which have major runways in the predominant strongest wind direction. The Lockheed C-5 Galaxy has swivelling 6-wheel main units for crosswind landings and castoring rear units to prevent tire scrubbing on tight turns.

## Kneeling Gear

One of the very first aircraft to use a "kneeling" function in its undercarriage design was the World War II German Arado Ar 232 cargo/transport aircraft, produced in small numbers as both a twin-engined version, and one with four engines - both the nosegear, and the wing-mounted, inwards-retracting main landing gear were designed to have a "kneeling" function in their design to assist in loading/unloading cargo, and to also allow its unique, exposed fixed ventral fuselage-centreline set of eleven "twinned" auxiliary wheel sets to more firmly support the fuselage on soft ground, and to enable taxiing the aircraft over ditches and other ground obstacles.

Some early U.S. Navy jet fighters were equipped with "kneeling" nose gear consisting of small steerable auxiliary wheels on short struts located forward of the primary nose gear, allowing the aircraft to be taxied tail-high with the primary nose gear retracted. This feature was intended to enhance safety aboard aircraft carriers by redirecting the hot exhaust blast upwards, and to reduce hangar space requirements by enabling the aircraft to park with its nose underneath the tail of a similarly equipped jet. Kneeling gear was used on the North American FJ-1 Fury and on early versions of the McDonnell F2H Banshee, but was found to be of little use operationally, and was omitted from later Navy fighters.

The nosewheel on the Lockheed C-5, partially retracts against a bumper to assist in loading and unloading of cargo using ramps through the forward, "tilt-up" hinged fuselage nose while stationary on the ground. The aircraft also tilts backwards. The Messier twin-wheel main units fitted to the Transall and other cargo aircraft can tilt forward or backward as necessary.

The Boeing AH-64 Apache helicopter is able to kneel to fit inside the cargo hold of a transport aircraft and for storage.

## Tail Support

Aircraft landing gear includes devices to prevent fuselage contact with the ground by tipping back when the aircraft is being loaded. Some commercial aircraft have used tail props when parked at the gate. The Douglas C-54 had a critical CG location which required a ground handling strut. The Lockheed C-130 and Boeing C-17 Globemaster III use ramp supports.

## Monowheel

A Schleicher ASG 29 glider shows its monowheel landing gear.

To minimize drag, modern gliders usually have a single wheel, retractable or fixed, centered under the fuselage, which is referred to as *monowheel gear* or *monowheel landing gear*. Monowheel gear is also used on some powered aircraft, where drag reduction is a priority, such as the Europa XS. Much like the Me 163 rocket fighter, some gliders from prior to the Second World War used a take-off dolly that was jettisoned on take-off and then landed on a fixed skid. This configuration is necessarily accompanied with a taildragger.

## Helicopters

Light helicopters use simple landing skids to save weight and cost. The skids may have attachment points for wheels so that they can be moved for short distances on the ground. Skids are impractical for helicopters weighing more than four tons. Some high-speed machines have retractable wheels, but most use fixed wheels for their robustness, and to avoid the need for a retraction mechanism.

## Tailsitter

A Convair XFY Pogo showing its landing gear.

Experimental tailsitter aircraft use landing gear located in their tails for VTOL operation.

## Light Aircraft

For light aircraft a type of landing gear which is economical to produce is a simple wooden arch laminated from ash, as used on some homebuilt aircraft. A similar arched gear is often formed from spring steel. The Cessna Airmaster was among the first aircraft to use spring steel landing gear. The main advantage of such gear is that no other shock-absorbing device is needed; the deflecting leaf provides the shock absorption.

## Folding Gear

Ju 288 V1 first prototype, showing its complex "folding" main undercarriage.

The limited space available to stow landing gear has led to many complex retraction mechanisms, each unique to a particular aircraft. An early example, the German *Bomber B* combat aircraft design competition winner, the Junkers Ju 288, had a complex "folding" main landing gear unlike any other aircraft designed by either Axis or Allied sides in the war: its single oleo strut was only attached to the lower end of its Y-form main retraction struts, handling the twinned main gear wheels, and folding by swiveling downwards and aftwards during retraction to "fold" the maingear's length to shorten it for stowage in the engine nacelle it was mounted in. However, the single pivot-point design also led to numerous incidents of collapsed maingear units for its prototype airframes.

## Tracked

Experimental tracked gear on a B-36 Peacemaker.

Increased contact area can be obtained with very large wheels, lots of smaller wheels or track-type gear. Tracked gear made by Dowty was fitted to a Westland Lysander in 1938 for taxi tests, then a Fairchild Cornell and a Douglas Boston. Bonmartini, in Italy, fitted tracked gear to a Piper Cub in 1951. Track-type gear was also tested using a C-47, C-82 and B-50. A much heavier aircraft, an XB-36, was made available for further tests, although there was no intention of using it on production aircraft. The stress on the runway was reduced to one third that of the B-36 four-wheel bogie.

## Ground Carriage

Ground carriage is a long term (after 2030) concept of flying without landing gear. It is one of many aviation technologies being proposed to reduce greenhouse gas emissions. Leaving the landing gear on the ground reduces weight and drag. Leaving it behind after take-off was done for a different reason, ie with military objectives, during World War II using the "dolly" and "trolley" arrangements of the German Me 163B rocket fighter and Arado Ar 234A prototype jet recon-bomber.

## Steering

There are several types of steering. Taildragger aircraft may be steered by rudder alone (depending upon the prop wash produced by the aircraft to turn it) with a freely pivoting tail wheel, or by a steering linkage with the tail wheel, or by *differential braking* (the use of independent brakes on opposite sides of the aircraft to turn the aircraft by slowing one side more sharply than the other). Aircraft with tricycle landing gear usually have a steering linkage with the nosewheel (especially in large aircraft), but some allow the nosewheel to pivot freely and use differential braking and/or the rudder to steer the aircraft, like the Cirrus SR22.

Some aircraft require that the pilot steer by using rudder pedals; others allow steering with the yoke or control stick. Some allow both. Still others have a separate control, called a *tiller*, used for steering on the ground exclusively.

## Rudder

When an aircraft is steered on the ground exclusively using the rudder, it needs a substantial airflow past the rudder, which can be generated either by the forward motion of the aircraft or by propeller slipstream. Rudder steering requires considerable practice to use effectively. Although it needs airflow past the rudder, it has the advantage of not needing any friction with the ground, which makes it useful for aircraft on water,snow or ice.

## Direct

Some aircraft link the yoke, control stick, or rudder directly to the wheel used for steering. Manipulating these controls turns the steering wheel (the nose wheel for tricycle landing gear, and the tail wheel for taildraggers). The connection may be a firm one in which any movement of the controls turns the steering wheel (and vice versa), or it may be a soft one in which a spring-like mechanism twists the steering wheel but does not force it to turn. The former provides positive steering but makes it easier to skid the steering wheel; the latter provides softer steering (making it easy to overcontrol) but reduces the probability of skidding. Aircraft with retractable gear may disable the steering mechanism wholly or partially when the gear is retracted.

The nose gear steering-wheel (tiller) is visible as a semi-circular wheel to the left of the yoke in this photo of a Boeing 727 cockpit.

## Differential Braking

Differential braking depends on asymmetric application of the brakes on the main gear wheels to turn the aircraft. For this, the aircraft must be equipped with separate controls for the right and left brakes (usually on the rudder pedals). The nose or tail wheel usually is not equipped with brakes. Differential braking requires considerable skill. In aircraft with several methods of steering that include differential braking, differential braking may be avoided because of the wear it puts on the braking mechanisms. Differential braking has the advantage of being largely independent of any movement or skidding of the nose or tailwheel.

## Tiller

A tiller in an aircraft is a small wheel or lever, sometimes accessible to one pilot and sometimes duplicated for both pilots, that controls the steering of the aircraft while it is on the ground. The tiller may be designed to work in combination with other controls such as the rudder or yoke. In large airliners, for example, the tiller is often used as the sole means of steering during taxi, and then the rudder is used to steer during takeoff and landing, so that both aerodynamic control surfaces and the landing gear can be controlled simultaneously when the aircraft is moving at aerodynamic speeds.

## Tires and Wheels

The specified selection criterion, e.g., minimum size, weight, or pressure, are used to select suitable tires and wheels from manufacturer's catalog and industry standards found in the Aircraft Yearbook published by the Tire and Rim Association, Inc.

Two mechanics replacing a main landing gear wheel on a Lockheed P-3 Orion.

## Gear Loading

The choice of the main wheel tires is made on the basis of the static loading case. The total main gear load $F_m$ is calculated assuming that the aircraft is taxiing at low speed without braking:

$$F_m = \frac{l_n}{l_m + l_n} W.$$

where $W$ is the weight of the aircraft and $l_m$ and $l_n$ are the distance measured from the aircraft's center of gravity(cg) to the main and nose gear, respectively.

The choice of the nose wheel tires is based on the nose wheel load $F_n$ during braking at maximum effort:

$$F_n = \frac{l_m}{l_m + l_n}(W - L) + \frac{h_{cg}}{l_m + l_n}\left(\frac{a_x}{g} W - D + T\right).$$

where $L$ is the lift, $D$ is the drag, $T$ is the thrust, and $h_{cg}$ is the height of aircraft cg from the static groundline. Typical values for $\frac{a_x}{g}$ on dry concrete vary from 0.35 for a simple brake system to 0.45 for an automatic brake pressure control system. As both $L$ and $D$ are positive, the maximum nose gear load occurs at low speed. Reverse thrust decreases the nose gear load and hence the condition $T = 0$ results in the maximum value:

$$F_n = \frac{l_m + h_{cg}(\frac{a_x}{g})}{l_m + l_n} W.$$

To ensure that the rated loads will not be exceeded in the static and braking conditions, a seven percent safety factor is used in the calculation of the applied loads.

## Inflation Pressure

Provided that the wheel load and configuration of the landing gear remain unchanged, the weight and volume of the tire will decrease with an increase in inflation pressure. From the flotation standpoint, a decrease in the tire contact area will induce a higher bearing stress on the pavement which may reduce the number of airfields available to the aircraft. Braking will also become less effective due to a reduction in the frictional force between the tires and the ground. In addition, the decrease in the size of the tire, and hence the size of the wheel, could pose a problem if internal brakes are to be fitted inside the wheel rims. The arguments against higher pressure are of such a nature that commercial operators generally prefer the lower pressures in order to maximize tire life and minimize runway stress. To prevent punctures from stones Philippine Airlines had to operate their Hawker Siddeley 748 aircraft with pressures as low as the tire manufacturer would permit. However, too low a pressure can lead to an accident as in the Nigeria Airways Flight 2120.

A rough general rule for required tire pressure is given by the manufacturer in their catalog. Goodyear for example advises the pressure to be 4% higher than required for a given weight or as fraction of the rated static load and inflation.

Tires of many commercial aircraft are required to be filled with nitrogen, and not subsequently diluted with more than 5% oxygen, to prevent auto-ignition of the gas which may result from overheating brakes producing volatile vapors from the tire lining.

Naval aircraft use different pressures when operating from a carrier and ashore. For example, the Northrop Grumman E-2 Hawkeye tire pressures are 260 psi on ship and 210 psi ashore. En-route deflation is used in tne Lockheed C-5 Galaxy to suit airfield conditions at the destination but adds excessive complication to the landing gear and wheels.

## Future Developments

Noise Airport community noise is an environmental issue which has brought into focus the contribution of aerodynamic noise from the landing gear. A NASA long term goal is to confine aircraft objectional noise to within the airport bounary. During the approach to land the landing gear is lowered several miles from touchdown and the landing gear is the dominant airframe noise source, followed by deployed highlift devices. With engines at a reduced power setting on the approach it is necessary to reduce airframe noise to make a significant reduction to total aircraft noise. The addition of add-on fairings is one approach for reducing the noise from the landing gear with a longer term approach to address noise generation during initial design.

Semi-active gear Airline specifications require an airliner to reach up to 90,000 take-offs and landings and roll 500,000 km on the ground in its lifetime. Conventional landing gear is designed to absorb the energy of a landing and doesn't perform well at reducing ground-induced vibrations in the airframe during landing ground roll, taxi and take-off. Airframe vibrations and fatigue damage can be reduced using semi-active oleos which vary damping over a wide range of ground speeds and runway quality.

## Accidents

Malfunctions or human errors (or a combination of these) related to retractable landing gear have been the cause of numerous accidents and incidents throughout aviation history. Distraction and preoccupation during the landing sequence played a prominent role in the approximately 100 gear-up landing incidents that occurred each year in the United States between 1998 and 2003. A gear-up landing, also known as a belly landing, is an accident that results from the pilot forgetting to lower the landing gear, or being unable to do so because of a malfunction. Although rarely fatal, a gear-up landing can be very expensive if it causes extensive airframe/engine damage. For propeller-driven aircraft a prop strike may require an engine overhaul.

JetBlue Airways Flight 292, an Airbus A320, making an emergency landing on runway 25L at Los Angeles International Airport in 2005 after the front landing gear malfunctioned.

Some aircraft have a stiffened fuselage underside or added features to minimize structural damage in a wheels-up landing. When the Cessna Skymaster was converted for a military spotting role (the O-2 Skymaster), fiberglass railings were added to the length of the fuselage; they were adequate to support the aircraft without damage if it was landed on a grassy surface.

The Bombardier Dash 8 is notorious for its landing gear problems. There were three incidents involved, all of them involving Scandinavian Airlines, flights SK1209, SK2478, and SK2867. This led to Scandinavian retiring all of its Dash 8s. The cause of these incidents was a locking mechanism that failed to work properly. This also caused concern for the aircraft for many other airlines that found similar problems, Bombardier Aerospace ordered all Dash 8s with 10,000 or more hours to be grounded, it was soon found that 19 Horizon Airlines Dash 8s had locking mechanism problems, so did 8 Austrian Airlines planes, this did cause several hundred flights to be canceled.

## Emergency Extension Systems

In the event of a failure of the aircraft's landing gear extension mechanism a backup is provided. This may be an alternate hydraulic system, a hand-crank, compressed air (nitrogen), pyrotechnic or free-fall system.

A free-fall or gravity drop system uses gravity to deploy the landing gear into the down and locked position. To accomplish this the pilot activates a switch or mechanical handle in the cockpit, which releases the up-lock. Gravity then pulls the landing gear down and deploys it. Once in position the landing gear is mechanically locked and safe to use for landing.

## Ground Resonance in Rotorcraft

Rotorcraft with fully articulated rotors may experience a dangerous and self-perpetuating phenomenon known as ground resonance, in which the unbalanced rotor system vibrates at a frequency coinciding with the natural frequency of the airframe, causing the entire aircraft to violently shake or wobble in contact with the ground. Ground resonance occurs when shock is continuously transmitted to the turning rotors through the landing gear, causing the angles between the rotor blades to become uneven; this is typically triggered if the aircraft touches the ground with forward or lateral motion, or touches down on one corner of the landing gear due to sloping ground or the craft's flight attitude. The resulting violent oscillations may cause the rotors or other parts to catastrophically fail, detach, and/or strike other parts of the airframe; this can destroy the aircraft in seconds and critically endanger persons unless the pilot immediately initiates a takeoff or closes the throttle and reduces rotor pitch. Ground resonance was cited in 34 National Transportation Safety Board incident and accident reports in the United States between 1990 and 2008.

Rotorcraft with fully articulated rotors typically have shock-absorbing landing gear designed to prevent ground resonance; however, poor landing gear maintenance and improperly inflated tires may contribute to the phenomenon. Helicopters with skid-type landing gear are less prone to ground resonance than those with wheels.

## Stowaways

Unauthorized passengers have been known to stowaway on larger aircraft by climbing a landing gear strut and riding in the compartment meant for the wheels. There are extreme dangers to this practice, with numerous deaths reported. Dangers include a lack of oxygen at high altitude, temperatures well below freezing, crush injury or death from the gear retracting into its confined space, and falling out of the compartment during takeoff or landing.

## Spacecraft

## Launch Vehicles

Landing gear has traditionally not been used on the vast majority of space launch vehicles, which take off vertically and are destroyed on falling back to earth. With some exceptions for suborbital vertical-landing vehicles (e.g., Masten Xoie or the Armadillo Aerospace' Lunar Lander Challenge vehicle), or for spaceplanes that use the *vertical takeoff, horizontal landing* (VTHL) approach (e.g., the Space Shuttle, or the USAF X-37), landing gear have been largely absent from orbital vehicles during the early decades since the advent of spaceflight technology, when orbital space transport has been the exclusive preserve of national-monopoly governmental space programs. Each spaceflight system through 2015 had relied on expendable boosters to begin each ascent to orbital velocity.

Falcon 9 descending, just after landing legs were extended.

Advances during the 2010s in private space transport, where new competition to governmental space initiatives has emerged, have included the explicit design of landing gear into orbital booster rockets. SpaceX has initiated and funded a multimillion-dollar reusable launch system development program to pursue this objective. As part of this program, SpaceX built, and flew eight times in 2012–2013, a first-generation booster-test-vehicle with a large fixed landing gear in order to test low-altitude vehicle dynamics and control for vertical landings of a near-empty orbital first stage. A second-generation larger booster test vehicle was built with extensible landing gear. The first prototype was flown four times—with all landing attempts successful—in 2014 for low-altitude tests before being self-destructed for safety reasons on a fifth test flight due to a blocked engine sensor port.

The orbital-flight version of the SpaceX design—being flown on both Falcon 9 and Falcon Heavy launch vehicles—includes a lightweight, deployable landing gear for the booster stage: a nested, telescoping piston on an A-frame. The total span of the four carbon fiber/aluminum extensible landing legs is approximately 18 metres (60 ft), and weigh less than 2,100 kilograms (4,600 lb); the deployment system uses high-pressure Helium as the working fluid. The first test of the extensible landing gear was successfully accomplished in April 2014 on a Falcon 9 rocket returning from an orbital launch and was the first successful controlled ocean soft touchdown of a liquid-rocket-engine orbital booster. After a single successful booster recovery in 2015, and several in 2016, the recovery of SpaceX booster stages became routine by 2017. Landing legs had become an ordinary operational part of orbital spaceflight launch vehicles.

SpaceX *Starhopper*, 9-meter-diameter test article.

The newest launch vehicle under development at SpaceX—the Starship—has landing legs on its first stage (Super Heavy) like Falcon 9 but also has three landing legs on its reusable second stage, a first for launch vehicle second stages.

Comet lander Philae showing landing gear. Each pad at the end of each of the lander legs has an ice screw, necessary for attachment to a celestial body with a very low gravitational field.

## Landers

Spacecraft designed to land safely on extraterrestrial bodies such as the Moon or Mars usually have landing gear. Such landers include the Apollo Lunar Module as well as a number of robotic space probe landers. Examples include Viking 1 lander, the first lander to successfully land on Mars (November 1976), and Philae which arrived at comet 67P/Churyumov–Gerasimenko in 2014 after a 10-year transit and landed on the comet on 12 November 2014.

Spacecraft with landing gear designed for use on non-terrestrial surfaces now under development include Prospector-1, with launch planned by 2020, and the large 85-tonne (187,000 lb) Starship, being developed for flight in the early 2020s.

# PROPELLERS

The propeller, the unit that must absorb the power output of the engine, has passed through many stages of development. Although most propellers are two-bladed, great increases in power output have resulted in the development of four and six-bladed propellers of large diameters. However, all propeller-driven aircraft are limited by the revolutions per minute (rpm) at which propellers can be turned.

There are several forces acting on the propeller as it turns; a major one is centrifugal force. This force at high rpm tends to pull the blades out of the hub, so blade weight is very important to the design of a propeller. Excessive blade tip speed (rotating the propeller too fast) may result not only in poor blade efficiency, but also in fluttering and vibration. Since the propeller speed is limited, the aircraft speed of a propeller driven aircraft is also limited—to approximately 400 miles per hour (mph). As aircraft speeds increased, turbofan engines were used for higher speed aircraft. Propeller-driven aircraft have several advantages and are widely used for applications in turboprops and reciprocating engine installations. Takeoff and landing can be shorter and less expensive. New blade materials and manufacturing techniques have increased the efficiency of propellers. Many smaller aircraft will continue to use propellers well into the future.

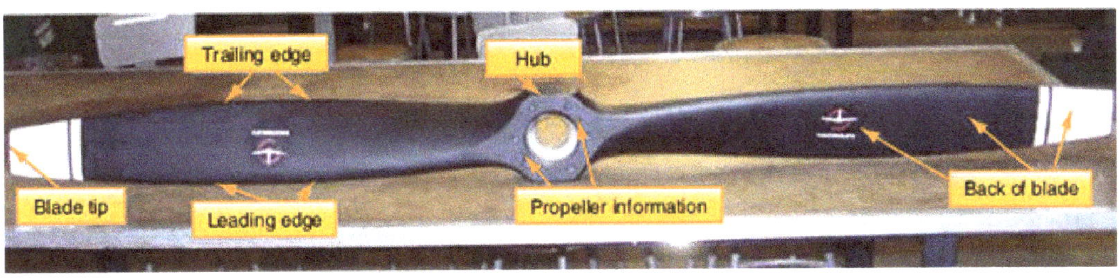


Basic nomenclature of propellers.

The basic nomenclature of the parts of a propeller is shown in Figure for a simple fixed-pitch, two-bladed wood propeller. The aerodynamic cross-section of a blade in Figure includes terminology to describe certain areas shown.

The many different types of propeller systems have been developed for specific aircraft installation, speed, and mission. Propeller development has encouraged many changes as propulsion systems have evolved. The first propellers were fabric-covered sticks made to force air in a rearward direction. Propellers started as simple two bladed wood propellers and have advanced to the complex propulsion systems of turboprop aircraft that involve more than just the propeller. As an outgrowth of operating large, more complex propellers, a variable-pitch, constant-speed feathering and reversing propeller system was developed. This system allows the engine rpm to be varied only slightly during different flight conditions and, therefore, increases flying efficiency.

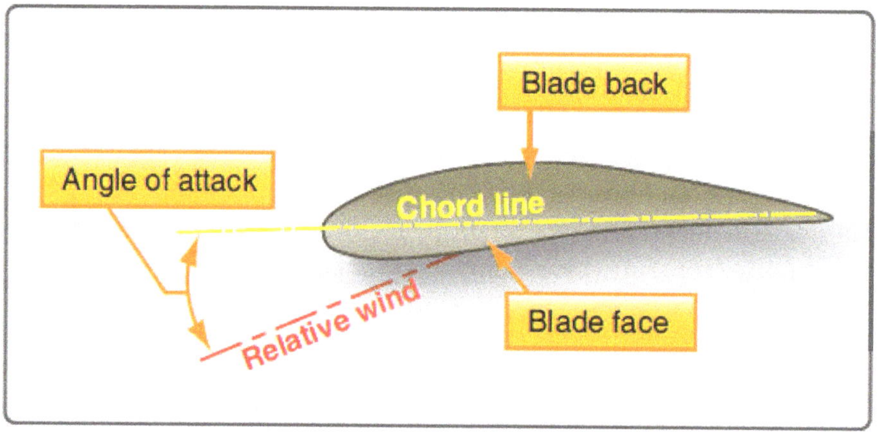


Cross-sectional area of a propeller blade airfoil.

A basic constant-speed system consists of a flyweight-equipped governor unit that controls the pitch angle of the blades so that the engine speed remains constant. The governor can be regulated by controls in the cockpit so that any desired blade angle setting and engine operating speed can be obtained. A low-pitch, high-rpm setting, for example, can be utilized for takeoff. Then, after the aircraft is airborne, a higher pitch and lower rpm setting can be used. Figure shows normal propeller movement with the positions of low pitch, high pitch, feather (used if the engine quits to reduce drag), and zero pitch into negative pitch, or reverse pitch.

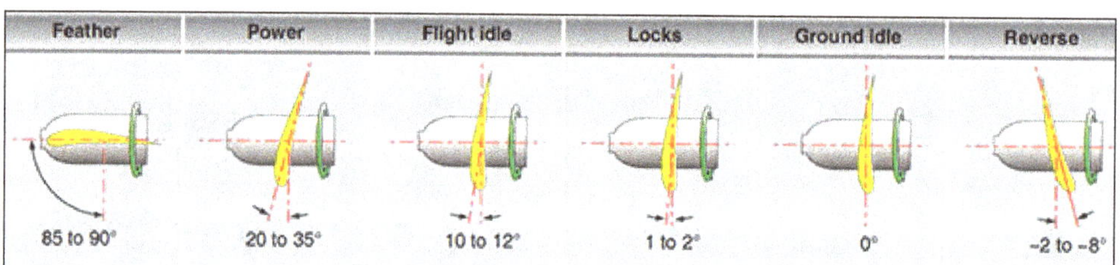


Typical propeller blade positions from feather position through the reverse position. Normal sequence of blade travel is feather, high pitch, low pitch, locks/ground idle, reverse pitch, and then back following the same path. Fixed turboprop engines are shut down on the locks to prevent load on the engine during restart.

Propeller range positions.

## Types of Propellers

There are various types or classes of propellers, the simplest of which are the fixed-pitch and ground-adjustable propellers. The complexity of propeller systems increases from these simpler forms to controllable-pitch and complex constant speed systems (automatic systems).

## Fixed-pitch Propeller

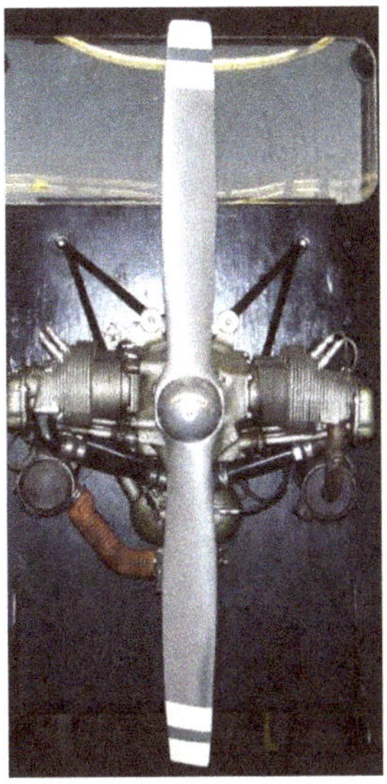

Fixed-pitch propeller.

As the name implies, a fixed-pitch propeller has the blade pitch, or blade angle, built into the propeller. The blade angle cannot be changed after the propeller is built. Generally, this type of propeller is one piece and is constructed of wood or aluminum alloy.

Fixed-pitch propellers are designed for best efficiency at one rotational and forward speed. They are designed to fit a set of conditions of both airplane and engine speeds and any change in these conditions reduces the efficiency of both the propeller and the engine. The fixed-pitch propeller is used on airplanes of low power, speed, range, or altitude. Many single-engine aircraft use fixed-pitch propellers and the advantages to these are less expense and their simple operation. This type of propeller does not require any control inputs from the pilot in flight.

## Test Club Propeller

*Test club.*

A test club is used to test and break in reciprocating engines. They are made to provide the correct amount of load on the engine during the test break-in period. The multi-blade design also provides extra cooling air flow during testing.

## Ground-adjustable Propeller

The ground-adjustable propeller operates as a fixed-pitch propeller. The pitch, or blade angle, can be changed only when the propeller is not turning. This is done by loosening the clamping mechanism that holds the blades in place. After the clamping mechanism has been tightened, the pitch of the blades cannot be changed in flight to meet variable flight requirements. The ground-adjustable propeller is not often used on present-day airplanes.

## Controllable-pitch Propeller

The controllable-pitch propeller permits a change of blade pitch, or angle, while the propeller is rotating. This allows the propeller to assume a blade angle that gives the best performance for particular flight conditions. The number of pitch positions may be limited, as with a two-position controllable propeller, or the pitch may be adjusted to any angle between the minimum and maximum pitch settings of a given propeller. The use of controllable-pitch propellers also makes it possible to attain the desired engine rpm for a particular flight condition.

This type of propeller is not to be confused with a constant speed propeller. With the controllable-pitch type, the blade angle can be changed in flight, but the pilot must change the propeller blade angle directly. The blade angle will not change again until the pilot changes it. The use of a governor is the next step in the evolution of propeller development, making way for constant-speed propellers with governor systems. An example of a two-position propeller is a Hamilton Standard counterweight two-position propeller. These types of propeller are not in wide use today.

## Constant-speed Propellers

The propeller has a natural tendency to slow down as the aircraft climbs and to speed up as the aircraft dives because the load on the engine varies. To provide an efficient propeller, the speed is kept as constant as possible. By using propeller governors to increase or decrease propeller pitch, the engine speed is held constant. When the airplane goes into a climb, the blade angle of the propeller decreases just enough to prevent the engine speed from decreasing. The engine can maintain its power output if the throttle setting is not changed. When the airplane goes into a dive, the blade angle increases sufficiently to prevent overspeeding and, with the same throttle setting, the power output remains unchanged. If the throttle setting is changed instead of changing the speed of the airplane by climbing or diving, the blade angle increases or decreases as required to maintain a constant engine rpm. The power output (not the rpm) changes in accordance with changes in the throttle setting. The governor-controlled, constant-speed propeller changes the blade angle automatically, keeping engine rpm constant.

One type of pitch-changing mechanism is operated by oil pressure (hydraulically) and uses a piston-and-cylinder arrangement. The piston may move in the cylinder, or the cylinder may move over a stationary piston. The linear motion of the piston is converted by several different types of mechanical linkage into the rotary motion necessary to change the blade angle. The mechanical connection may be through gears, the pitch-changing mechanism that turns the butt of each blade. Each blade is mounted with a bearing that allows the blade to rotate to change pitch. In most cases, the oil pressure for operating the different types of hydraulic pitch-changing mechanisms comes directly from the engine lubricating system. When the engine lubricating system is used, the engine oil pressure is usually boosted by a pump that is integral with the governor to operate the propeller. The higher oil pressure (approximately 300 pounds per square inch (psi)) provides a quicker blade-angle change. The governors direct the pressurized oil for operation of the hydraulic pitch-changing mechanisms.

Blade bearing areas in hub.

The governors used to control hydraulic pitch-changing mechanisms are geared to the engine crankshaft and are sensitive to changes in rpm. When rpm increases above the value for which a governor is set, the governor causes the propeller pitch-changing mechanism to turn the blades to a higher angle. This angle increases the load on the engine, and rpm decreases. When rpm decreases below the value for which a governor is set, the governor causes the pitch changing mechanism to turn the blades to a lower angle; the load on the engine is decreased, and rpm increases. Thus, a propeller governor tends to keep engine rpm constant.

In constant-speed propeller systems, the control system adjusts pitch through the use of a governor, without attention by the pilot, to maintain a specific preset engine rpm within the set range of the propeller. For example, if engine speed increases, an overspeed condition occurs and the propeller needs to slow down. The controls automatically increase the blade angle until desired rpm has been reestablished. A good constant speed control system responds to such small variations of rpm that for all practical purposes, a constant rpm is maintained.

Each constant-speed propeller has an opposing force that operates against the oil pressure from the governor. Flyweights mounted to the blades move the blades in the high pitch direction as the propeller turns. Other forces used to move the blades toward the high pitch direction include air pressure (contained in the front dome), springs, and aerodynamic twisting moment.

## Feathering Propellers

Feathering propellers must be used on multi-engine aircraft to reduce propeller drag to a minimum under one or more engine failure conditions. A feathering propeller is a constant-speed propeller used on multi-engine aircraft that has a mechanism to change the pitch to an angle of approximately 90°. A propeller is usually feathered when the engine fails to develop power to turn the propeller. By rotating the propeller blade angle parallel to the line of flight, the drag on the aircraft is greatly reduced. With the blades parallel to the airstream, the propeller stops turning and minimum windmilling, if any, occurs. The blades are held in feather by aerodynamic forces.

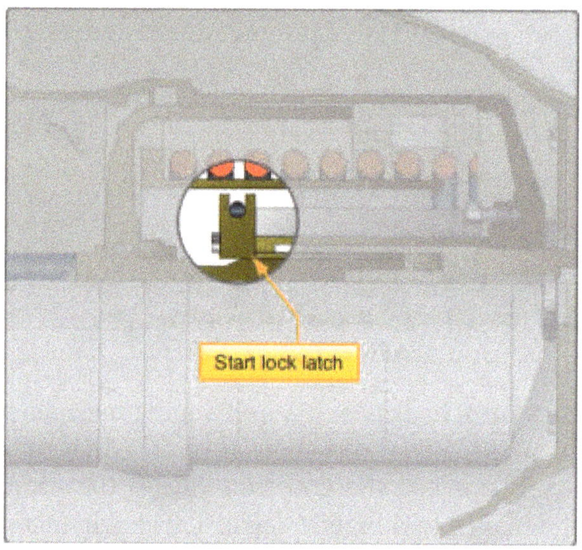


Feathering latches.

Almost all small feathering propellers use oil pressure to take the propeller to low pitch and blade flyweights, springs, and compressed air to take the blades to high pitch. Since the blades would go to the feather position during shutdown, latches lock the propeller in the low pitch position as the propeller slows down at shutdown. These can be internal or external and are contained within the propeller hub. In flight, the latches are prevented from stopping the blades from feathering because they are held off their seat by centrifugal force. Latches are needed to prevent excess load on the engine at start up. If the blade were in the feathered position during engine start, the engine would be placed under an undue load during a time when the engine is already subject to wear.

## Reverse-pitch Propellers

Additional refinements, such as reverse-pitch propellers (mainly used on turbo props), are included in some propellers to improve their operational characteristics. Almost all reverse-pitch propellers are of the feathering type. A reverse pitch propeller is a controllable propeller in which the blade angles can be changed to a negative value during operation.

The purpose of the reversible pitch feature is to produce a negative blade angle that produces thrust opposite the normal forward direction. Normally, when the landing gear is in contact with the runway after landing, the propellers blades can be moved to negative pitch (reversed), which creates thrust opposite of the aircraft direction and slows the aircraft. As the propeller blades move into negative pitch, engine power is applied to increase the negative thrust. This aerodynamically brakes the aircraft and reduces ground roll after landing. Reversing the propellers also reduces aircraft speed quickly on the runway just after touchdown and minimizes aircraft brake wear.

## References


- Basic-construction: home.iitk.ac.in, Retrieved 13 June, 2020
- Propellers-2: flight-mechanic.com, Retrieved 23 April, 2020
- E. V. Laitone, Wind tunnel tests of wings at Reynolds numbers below 70 000, Experiments in Fluids 23, 405 (1997). doi:10.1007/s003480050128
- Aviation Publishers Co. Limited, From the Ground Up, p. 14 (27th revised edition) ISBN 0-9690054-9-0
- Types-of-propellers: flight-mechanic.com, Retrieved 18 July, 2020

4

# Aircraft Engines

Aircraft engine is the major component of the aircraft that helps in generating mechanical power. In-line engine, V- type engine, radial engine, Wankel engine, gas turbine, pulse jets, scramjet, turboshaft, etc. are some of the types of engines. This chapter sheds light on the subject of these types of aircraft engines for providing an in-depth understanding of it.

## RECIPROCATING OR PISTON ENGINE

An aircraft piston engine, also commonly referred to as a reciprocating engine or "recip", is an internal combustion engine that uses one or more reciprocating pistons to convert pressure into a rotational motion. The aircraft piston engine operates on the same principles as the engines found in most automobiles. However, modifications, such as dual ignition systems, to improve redundancy and safety, and air cooling to reduce weight, have been incorporated into engines designed for aviation use. Turbochargers and, less commonly, superchargers can be added to piston engines to improve performance. Aircraft piston engines are most commonly fueled with AVGAS but diesel fueled engines are becoming more common, especially in light aircraft.

### In-line Engine

A Mercedes D.II inline engine on display.

In aviation, an inline engine is a reciprocating engine with banks of cylinders, one behind another, rather than rows of cylinders, with each bank having any number of cylinders, but rarely more than six.

## Inverted Engine

Argus As 10 inverted air-cooled V8 engine.

Some inline aircraft engines, including the de Havilland Gipsy Major used in the de Havilland Tiger Moth, and the Daimler-Benz DB 601 used in the Messerschmitt Bf 109, were designed to be installed in airframes inverted, such that the crankshaft was at the top of the engine and the cylinder heads were at the bottom. The advantages of inverted engines included improved visibility for the pilot, improved access to cylinder heads and manifolds for the ground crew, having the centre of mass of a multi-bank engine lower in the engine, and having the widest part of a multi-bank engine being closer to the midline of the fuselage, which is also generally wide.

## V-type Engine

An early Vee engine - this is a two-cylinder V-twin used in an early British motorcycle.

A V engine, or Vee engine is a common configuration for an internal combustion engine. The cylinders and pistons are aligned, in two separate planes or 'banks', so that they appear to be in a "V" when viewed along the axis of the crankshaft. The Vee configuration generally reduces the overall engine length, height and weight compared with an equivalent inline configuration.

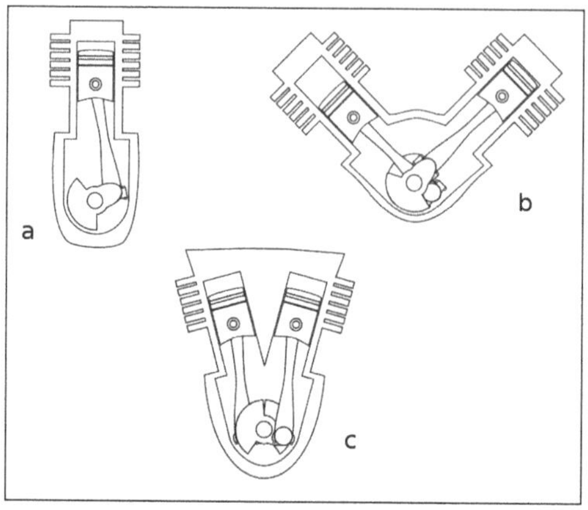


Three types of engine: a — straight engine, b — Vee engine, c — VR engine.

## Characteristics

Usually, each pair of corresponding pistons from each bank of cylinders share one crankpin on the crankshaft, either by master/slave rods or by two ordinary connecting rods side by side. However, some V-twin engine designs have two-pin cranks, while other V configurations include split crank-pins for more even firing.

V-engines are generally more compact than straight engines with cylinders of the same dimensions and number. This effect increases with the number of cylinders in the engine; there might be no noticeable difference in overall size between V-twin and straight-twin engines while V8 engines are much more compact than straight-eight engines.

Various cylinder bank angles of Vee are used in different engines; depending on the number of cylinders, there may be angles that work better than others for stability. Very narrow angles of Vee combine some of the advantages of the Vee engine and the straight engine (primarily in the form of compactness) as well as disadvantages; the concept is an old one pioneered by Lancia's V4 engine in the 1920s, but recently reworked by Volkswagen Group with their VR engines, which is actually a combination of V and straight configuration.

Some Vee configurations are well-balanced and smooth, while others are less smoothly running than their equivalent straight counterparts. V8s with crossplane crankshaft can be easily balanced with the use of counterweights only. V12s, being in effect two straight-6 engines married together, are fully balanced; if the V-angle is 60° for 4-stroke or 30° for 2-stroke, they also have even firing. Others, such as the V2, V4, V6, flatplane V8, V10, V14 and V18 engine show increased vibration and generally require balance shafts or split crankshafts.

Certain types of Vee engine have been built as inverted engines, most commonly for aircraft. Advantages include better visibility in a single-engined airplane, and lower centre of gravity. Examples include World War II German Daimler-Benz DB 601, Junkers Jumo, and Argus Motoren piston engines.

## Horizontally Opposed Engine

Also known as flat or boxer engines, these aircraft piston engines have two banks of cylinders on opposite sides of a central crankcase. These engines can be air or liquid cooled, but are most often air cooled. Reliability, simplicity and easy maintenance have made this type of piston engine the most popular aircraft engine for more than half a decade. In addition to their efficiency, the engine can run on many different types of fuels as well. Older aircraft piston engines such as in-line, rotary or v-type need special care.

## Radial Engine

Radial engine of a biplane.

The radial engine is a reciprocating type internal combustion engine configuration in which the cylinders "radiate" outward from a central crankcase like the spokes of a wheel. It resembles a stylized star when viewed from the front, and is called a "star engine". The radial configuration was commonly used for aircraft engines before gas turbine engines became predominant.

Master rod (upright), slaves and balances from a two-row, fourteen-cylinder Pratt & Whitney Twin Wasp.

## Engine Operation

Mounting of a Wright R-2600 *Twin Cyclone* engine on a North American B-25 *Mitchell* bomber.

Since the axes of the cylinders are coplanar, the connecting rods cannot all be directly attached to the crankshaft unless mechanically complex forked connecting rods are used, none of which have been successful. Instead, the pistons are connected to the crankshaft with a master-and-articulating-rod assembly. One piston, the uppermost one in the animation, has a master rod with a direct attachment to the crankshaft. The remaining pistons pin their connecting rods' attachments to rings around the edge of the master rod. Extra "rows" of radial cylinders can be added in order to increase the capacity of the engine without adding to its diameter.

Four-stroke radials have an odd number of cylinders per row, so that a consistent every-other-piston firing order can be maintained, providing smooth operation. For example, on a five-cylinder engine the firing order is 1, 3, 5, 2, 4, and back to cylinder 1. Moreover, this always leaves a one-piston gap between the piston on its combustion stroke and the piston on compression. The active stroke directly helps compress the next cylinder to fire, making the motion more uniform. If an even number of cylinders were used, an equally timed firing cycle would not be feasible. The prototype radial Zoche aero-diesels have an even number of cylinders, either four or eight; but this is not problematic, because they are two-stroke engines, with twice the number of power strokes as a four-stroke engine per crankshaft rotation.

The radial engine normally uses fewer cam lobes than other types. As with most four-strokes, the crankshaft takes two revolutions to complete the four strokes of each piston (intake, compression, combustion, exhaust). The camshaft ring is geared to spin slower and in the opposite direction to the crankshaft. The cam lobes are placed in two rows for the intake and exhaust. For example, four cam lobes serve all five cylinders, whereas 10 would be required for a typical inline engine with the same number of cylinders and valves.

Most radial engines use overhead poppet valves driven by pushrods and lifters on a cam plate which is concentric with the crankshaft, with a few smaller radials, like the

Kinner B-5 and Russian Shvetsov M-11, using individual camshafts within the crankcase for each cylinder. A few engines use sleeve valves such as the 14-cylinder Bristol Hercules and the 18-cylinder Bristol Centaurus, which are quieter and smoother running but require much tighter manufacturing tolerances.

## Comparison with Inline Engines

The 1935 Monaco-Trossi race car, a rare example of automobile use.

Liquid cooling systems are generally more vulnerable to battle damage. Even minor shrapnel damage can easily result in a loss of coolant and consequent engine overheating, while an air-cooled radial engine may be largely unaffected by minor damage.Radials have shorter and stiffer crankshafts, a single-bank radial engine needing only two crankshaft bearings as opposed to the seven required for a liquid-cooled, six-cylinder, inline engine of similar stiffness.

While a single-bank radial permits all cylinders to be cooled equally, the same is not true for multi-row engines where the rear cylinders can be affected by the heat coming off the front row, and air flow being masked.

A potential disadvantage of radial engines is that having the cylinders exposed to the airflow increases drag considerably. The answer was the addition of specially designed cowlings with baffles to force the air between the cylinders. The first effective drag-reducing cowling that didn't impair engine cooling was the British Townend ring or "drag ring" which formed a narrow band around the engine covering the cylinder heads, reducing drag. The National Advisory Committee for Aeronautics studied the problem, developing the NACA cowling which further reduced drag and improved cooling. Nearly all aircraft radial engines since have used NACA-type cowlings.

While inline liquid-cooled engines continued to be common in new designs until late in World War II, radial engines dominated afterwards until overtaken by jet engines, with the late-war Hawker Sea Fury and Grumman F8F Bearcat, two of the fastest production piston-engined aircraft ever built, using radial engines.

## Other Types of Radial Engine

### Multi-row Radials

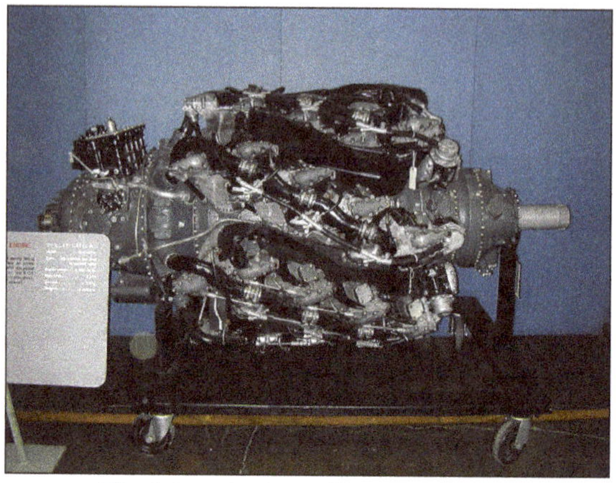

The Wasp Major, a four-row radial.

Originally radial engines had one row of cylinders, but as engine sizes increased it became necessary to add extra rows. The first radial-configuration engine known to use a twin-row design was the 160 hp Gnome "Double Lambda" rotary engine of 1912, designed as a 14-cylinder twin-row version of the firm's 80 hp Lambda single-row seven-cylinder rotary, however reliability and cooling problems limited its success.

Two-row designs began to appear in large numbers during the 1930s, when aircraft size and weight grew to the point where single-row engines of the required power were simply too large to be practical. Two-row designs often had cooling problems with the rear bank of cylinders, but a variety of baffles and fins were introduced that largely eliminated these problems. The downside was a relatively large frontal area that had to be left open to provide enough airflow, which increased drag. This led to significant arguments in the industry in the late 1930s about the possibility of using radials for high-speed aircraft like modern fighters.

The solution was introduced with the BMW 801 14-cylinder twin-row radial. Kurt Tank designed a new cooling system for this engine that used a high-speed fan to blow compressed air into channels that carry air to the middle of the banks, where a series of baffles directed the air over all of the cylinders. This allowed the cowling to be tightly fitted around the engine, reducing drag, while still providing (after a number of experiments and modifications) enough cooling air to the rear. This basic concept was soon copied by many other manufacturers, and many late-WWII aircraft returned to the radial design as newer and much larger designs began to be introduced. Examples include the Bristol Centaurus in the Hawker Sea Fury, and the Shvetsov ASh-82 in the Lavochkin La-7.

For even greater power, adding further rows was not considered viable due to the difficulty of providing the required airflow to the rear banks. Larger engines were designed, mostly using water cooling although this greatly increased complexity and eliminated some of the advantages of the radial air-cooled design. One example of this concept is the BMW 803, which never entered service.

A major study into the airflow around radials using wind tunnels and other systems was carried out in the US, and demonstrated that ample airflow was available with careful design. This led to the R-4360, which has 28 cylinders arranged in a 4 row *corncob* configuration. The R-4360 saw service on large American aircraft in the post-World War II period. The US and Soviet Union continued experiments with larger radials, but the UK abandoned such designs in favour of newer versions of the Centaurus and rapid movement to the use of turboprops such as the Armstrong Siddeley Python and Bristol Proteus, which easily produced more power than radials without the weight or complexity.

Large radials continued to be built for other uses, although they are no longer common. An example is the 5-ton Zvezda M503 diesel engine with 42 cylinders in 6 rows of 7, displacing 143.6 litres (8,760 cu in) and producing 3,942 hp (2,940 kW). Three of these were used on the fast Osa class missile boats.

## Diesel Radials

While most radial engines have been produced for gasoline, there have been diesel radial engines. Two major advantages favour diesel engines — lower fuel consumption and reduced fire risk.

## Packard

Packard DR-980 diesel radial aircraft engine.

Packard designed and built a 9-cylinder 980 cubic inch (16.06 litre) displacement diesel radial aircraft engine, the 225 horsepower (168 kW) DR-980, in 1928. On 28 May 1931, a DR-980 powered Bellanca CH-300, with 481 gallons of fuel, piloted by Walter Edwin Lees and Frederick Brossy set a record for staying aloft for 84 hours and

32 minutes without being refueled. This record stood for 55 years until broken by the Rutan Voyager.

## Bristol

The experimental Bristol Phoenix of 1928–1932 was successfully flight tested in a Westland Wapiti and set altitude records in 1934 that lasted until World War II.

## Clerget

In 1932 the French company Clerget developed the 14D, a 14-cylinder two-stroke diesel radial engine. After a series of improvements, in 1938 the 14F2 model produced 520 hp (390 kW) at 1910 rpm cruise power, with a power-to-weight ratio near that of contemporary gasoline engines and a specific fuel consumption of roughly 80% that for an equivalent gasoline engine. During WWII the research continued, but no mass-production occurred because of the Nazi occupation. By 1943 the engine had grown to produce over 1,000 hp (750 kW) with a turbocharger. After the war, the Clerget company was integrated in the SNECMA company and had plans for a 32-cylinder diesel engine of 4,000 hp (3,000 kW), but in 1947 the company abandoned piston engine development in favour of the emerging turbine engines.

## Nordberg

A Nordberg Manufacturing Company two-stroke diesel radial engine for power generation and pump drive purposes.

The Nordberg Manufacturing Company of the United States developed and produced a series of large two-stroke radial diesel engines from the late 1940s for electrical production, primarily at aluminum smelters and for pumping water. They differed from most radials in that they had an even number of cylinders in a single bank (or row) and an unusual double master connecting rod. Variants were built that could be run on either diesel oil or gasoline or mixtures of both. A number of powerhouse installations utilising large numbers of these engines were made in the U.S.

## EMD

Electro-Motive Diesel (EMD) built the "pancake" engines 16-184 and 16-338 for marine use.

## Compressed Air Radial Engines

A number of radial motors operating on compressed air have been designed, mostly for use in model airplanes and in gas compressors.

## Model Radial Engines

A number of multi-cylinder 4-stroke model engines have been commercially available in a radial configuration, beginning with the Japanese O.S. Max firm's FR5-300 five-cylinder, 3.0 cu.in. (50 $cm^3$) displacement "Sirius" radial in 1986. The American "Technopower" firm had made smaller-displacement five- and seven-cylinder model radial engines as early as 1976, but the OS firm's engine was the first mass-produced radial engine design in aeromodelling history. The rival Saito Seisakusho firm in Japan has since produced a similarly sized five-cylinder radial four-stroke model engine of their own as a direct rival to the OS design, with Saito also creating a series of three-cylinder methanol and gasoline-fueled model radial engines ranging from 0.90 cu.in. (15 $cm^3$) to 4.50 cu.in. (75 $cm^3$) in displacement, also all now available in spark-ignition format up to 84 $cm^3$ displacement for use with gasoline.

## Rotary Engine

This Le Rhone 9C installed on a Sopwith Pup fighter aircraft at the Fleet Air Arm Museum. The narrowness of the mounting pedestal to the fixed crankshaft, and the size of the engine.

The rotary engine was an early type of internal combustion engine, usually designed with an odd number of cylinders per row in a radial configuration, in which the crankshaft remained stationary in operation, with the entire crankcase and its attached cylinders rotating around it as a unit. Its main application was in aviation, although it also saw use before its primary aviation role, in a few early motorcycles and automobiles.

An 80 horsepower (60 kW) rated Le Rhone 9C, a typical rotary engine of WWI. The copper pipes carry the fuel-air mixture from the crankcase to the cylinder heads acting collectively as an intake manifold.

This type of engine was widely used as an alternative to conventional inline engines (straight or V) during World War I and the years immediately preceding that conflict. It has been described as "a very efficient solution to the problems of power output, weight, and reliability".

## Distinction between Rotary and Radial Engines

A rotary engine is essentially a standard Otto cycle engine, with cylinders arranged radially around a central crankshaft just like a conventional radial engine, but instead of having a fixed cylinder block with rotating crankshaft as with a radial engine, the crankshaft remains stationary and the entire cylinder block rotates around it. In the most common form, the crankshaft was fixed solidly to the airframe, and the propeller was simply bolted to the front of the crankcase.

This difference also has much impact on design (lubrication, ignition, fuel admission, cooling, etc.) and functioning.

The Musée de l'Air et de l'Espace in Paris has on display a special, "sectioned" working model of an engine with seven radially disposed cylinders. It alternates between rotary and radial modes to demonstrate the difference between the internal motions of the two types of engine.

## Arrangement

Like "fixed" radial engines, rotaries were generally built with an odd number of cylinders (usually 5, 7 or 9), so that a consistent every-other-piston firing order could be maintained, to provide smooth running. Rotary engines with an even number of cylinders were mostly of the "two row" type.

Most rotary engines were arranged with the cylinders pointing outwards from a single crankshaft, in the same general form as a radial, but there were also rotary boxer engines and even one-cylinder rotaries.

## Advantages and Drawbacks

Three key factors contributed to the rotary engine's success at the time:

- Smooth running: Rotaries delivered power very smoothly because (relative to the engine mounting point) there are no reciprocating parts, and the relatively large rotating mass of the crankcase/cylinders (as a unit) acted as a flywheel.
- Improved cooling: When the engine was running, the rotating crankcase/cylinder assembly created its own fast-moving cooling airflow, even with the aircraft at rest.
- Weight advantage: Many conventional engines had to have heavy flywheels added to smooth out power impulses and reduce vibration. Rotary engines gained a substantial power-to-weight ratio advantage by having no need for an added flywheel. They shared with other radial configuration engines the advantage of a small, flat crankcase, and because of their efficient air-cooling system, cylinders could be made with thinner walls and shallower cooling fins, which further reduced their weight.

Engine designers had always been aware of the many limitations of the rotary engine so when the static style engines became more reliable and gave better specific weights and fuel consumption, the days of the rotary engine were numbered.

- Rotary engines had a fundamentally inefficient total-loss oiling system. In order to reach the whole engine, the lubricating medium needed to enter the crankcase through the hollow crankshaft; but the centrifugal force of the revolving crankcase was directly opposed to any re-circulation. The only practical solution was for the lubricant to be aspirated with the fuel/air mixture, as in a two-stroke engine.
- Power increase also came with mass and size increases, multiplying gyroscopic precession from the rotating mass of the engine. This produced stability and control problems in aircraft in which these engines were installed, especially for inexperienced pilots.
- Power output increasingly went into overcoming the air-resistance of the spinning engine.
- Engine controls were tricky, and resulted in fuel waste.

## Rotary Engine Control

### Monosoupape Rotaries

It is often asserted that rotary engines had no throttle and hence power could only be reduced by intermittently cutting the ignition using a "blip" switch. This was almost literally true of the "Monosoupape" (single valve) type, which took most of the air into the cylinder through the exhaust valve, which remained open for a portion of the downstroke of the piston. Thus the richness of the mixture in the cylinder could not be controlled via the crankcase intake. The "throttle" (fuel valve) of a monosoupape provided only a very limited degree of speed regulation, as opening it made the mixture too rich, while closing it made it too lean (in either case quickly stalling the engine, or damaging the cylinders). Early models featured a pioneering form of variable valve timing in an attempt to give greater control, but this caused the valves to burn and therefore it was abandoned.

The only way of running a Monosoupape engine smoothly at reduced revs was with a switch that changed the normal firing sequence so that each cylinder fired only once per two or three engine revolutions, but the engine remained more or less in balance. As with excessive use of the "blip" switch: running the engine on such a setting for too long resulted in large quantities of unburned fuel and oil in the exhaust, and gathering in the lower cowling, where it was a notorious fire hazard.

## Normal Rotaries

Most rotaries had normal inlet valves, so that the fuel (and lubricating oil) was taken into the cylinders already mixed with air - as in a normal four-stroke engine. Although a conventional carburetor, with the ability to keep the fuel/air ratio constant over a range of throttle openings, was precluded by the spinning crankcase; it was possible to adjust the air supply through a separate flap valve or "bloctube". The pilot needed to set the throttle to the desired setting (usually full open) and then adjust the fuel/air mixture to suit using a separate "fine adjustment" lever that controlled the air supply valve (in the manner of a manual choke control). Due to the rotary engine's large rotational inertia, it was possible to adjust the appropriate fuel/air mixture by trial and error without stalling it, although this varied between different types of engine, and in any case it required a good deal of practice to acquire the necessary knack. After starting the engine with a known setting that allowed it to idle, the air valve was opened until maximum engine speed was obtained.

Throttling a running engine back to reduce revs was possible by closing off the fuel valve to the required position while re-adjusting the fuel/air mixture to suit. This process was also tricky, so that reducing power, especially when landing, was often accomplished instead by intermittently cutting the ignition using the blip switch.

Cutting cylinders using ignition switches had the drawback of letting fuel continue to pass through the engine, oiling up the spark plugs and making smooth restarting problematic. Also, the raw oil-fuel mix could collect in the cowling. As this could cause a serious fire when the switch was released, it became common practice for part or all of the bottom of the basically circular cowling on most rotary engines to be cut away, or fitted with drainage slots.

By 1918 a Clerget handbook advised maintaining all necessary control by using the fuel and air controls, and starting and stopping the engine by turning the fuel on and off. The recommended landing procedure involved shutting off the fuel using the fuel lever, while leaving the blip switch on. The windmilling propeller made the engine continue to spin without delivering any power as the aircraft descended. It was important to leave the ignition on to allow the spark plugs to continue to spark and keep them from oiling up, so that the engine could (if all went well) be restarted simply by re-opening the fuel valve. Pilots were advised to not use an ignition cut out switch, as it would eventually damage the engine.

Pilots of surviving or reproduction aircraft fitted with rotary engines still find that the blip switch is useful while landing, as it provides a more reliable, quicker way to initiate power if needed, rather than risk a sudden engine stall, or the failure of a windmilling engine to restart at the worst possible moment.

# TURBINE-POWERED

## Turboprop

GE T64 propeller at left, gearbox with accessories in the middle, turbine core at right.

A turboprop engine is a turbine engine that drives an aircraft propeller.

In its simplest form a turboprop consists of an intake, compressor, combustor, turbine, and a propelling nozzle. Air is drawn into the intake and compressed by the compressor. Fuel is then added to the compressed air in the combustor, where the fuel-air mixture then combusts. The hot combustion gases expand through the turbine. Some of the power generated by the turbine is used to drive the compressor. Thrust is obtained by the combusting gases, pushing toward a (vectored) surface in front of the expanding gas.

The rest is transmitted through the reduction gearing to the propeller. Further expansion of the gases occurs in the propelling nozzle, where the gases exhaust to atmospheric pressure. The propelling nozzle provides a relatively small proportion of the thrust generated by a turboprop.

In contrast to a turbojet, the engine's exhaust gases do not generally contain enough energy to create significant thrust, since almost all of the engine's power is used to drive the propeller.

## Technological Aspects

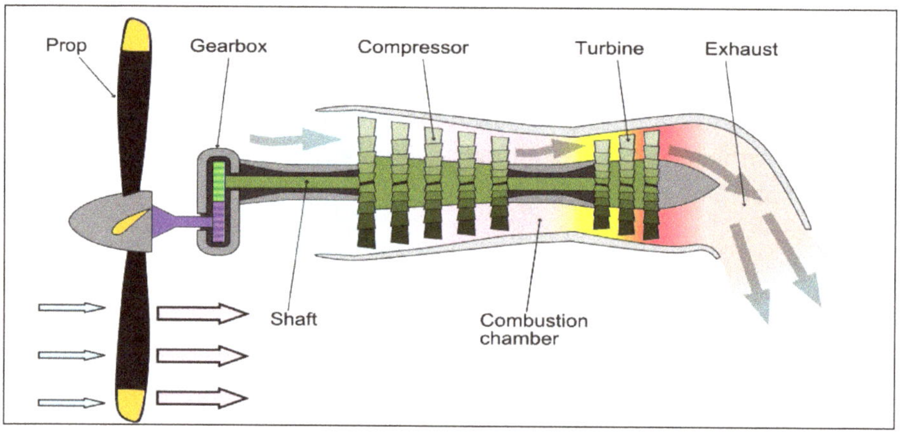


The operation of a turboprop engine.

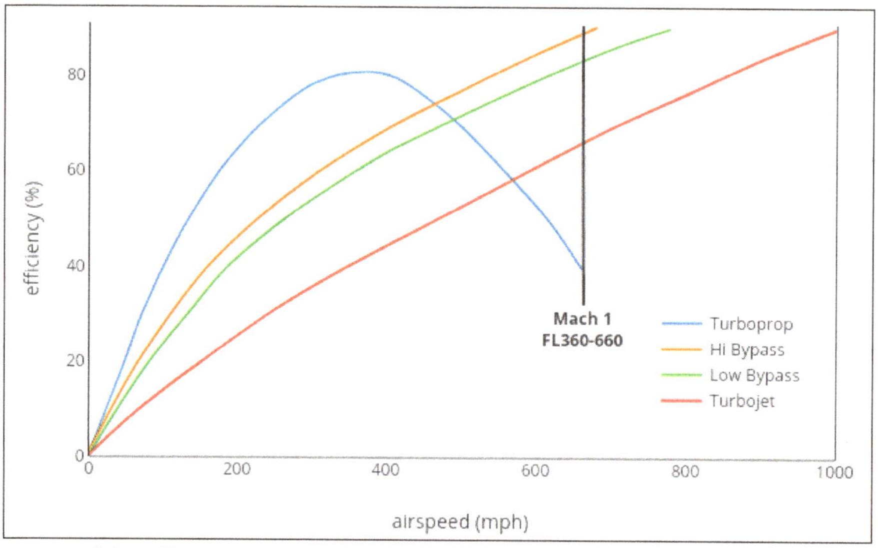


Propulsive efficiency comparison for various gas turbine engine configurations.

Exhaust thrust in a turboprop is sacrificed in favour of shaft power, which is obtained by extracting additional power (up to that necessary to drive the compressor) from turbine expansion. Owing to the additional expansion in the turbine system, the residual energy in the exhaust jet is low. Consequently, the exhaust jet typically produces around or less than 10% of the total thrust. A higher proportion of the thrust comes from the propeller at low speeds and less at higher speeds.

Turboprops can have bypass ratios up to 50-100 although the propulsion airflow is less clearly defined for propellers than for fans.

The propeller is coupled to the turbine through a reduction gear that converts the high RPM/low torque output to low RPM/high torque. The propeller itself is normally a constant-speed (variable pitch) propeller type similar to that used with larger aircraft reciprocating engines.

Unlike the small diameter fans used in turbofan jet engines, the propeller has a large diameter that lets it accelerate a large volume of air. This permits a lower airstream velocity for a given amount of thrust. As it is more efficient at low speeds to accelerate a large amount of air by a small degree than a small amount of air by a large degree, a low disc loading (thrust per disc area) increases the aircraft's energy efficiency, and this reduces the fuel use.

Propellers lose efficiency as aircraft speed increases, so turboprops are normally not used on high-speed aircraft above 0.6-0.7 Mach. However, propfan engines, which are very similar to turboprop engines, can cruise at flight speeds approaching 0.75 Mach. To increase propeller efficiency across a wider range of airspeeds, turboprops are typically equipped with constant-speed (variable-pitch) propellers. The blades of a constant-speed propeller increase pitch as aircraft speed increases, allowing for a wider range airspeeds than a fixed-pitch propeller. Another benefit of this type of propeller is that it can also be used to generate negative thrust while decelerating on the runway. Additionally, in the event of an engine failure, the propeller can be feathered, thus minimizing the drag of the non-functioning propeller.

While most modern turbojet and turbofan engines use axial-flow compressors, turboprop engines usually contain at least one stage of centrifugal compressor which have the advantage of being simple and lightweight, at the expense of a streamlined shape.

While the power turbine may be integral with the gas generator section, many turboprops today feature a free power turbine on a separate coaxial shaft. This enables the propeller to rotate freely, independent of compressor speed. Residual thrust on a turboshaft is avoided by further expansion in the turbine system and/or truncating and turning the exhaust 180 degrees, to produce two opposing jets. Apart from the above, there is very little difference between a turboprop and a turboshaft.

## Usage

Compared to turbofans, turboprops are most efficient at flight speeds below 725 km/h (450 mph; 390 knots) because the jet velocity of the propeller (and exhaust) is relatively low. Modern turboprop airliners operate at nearly the same speed as small regional jet airliners but burn two-thirds of the fuel per passenger. However, compared to a turbojet (which can fly at high altitude for enhanced speed and fuel efficiency) a propeller aircraft has a lower ceiling.

Compared to piston engines, their greater power-to-weight ratio (which allows for shorter takeoffs) and reliability can offset their higher initial cost, maintenance and fuel consumption. As jet fuel can be easier to obtain than avgas in remote areas, turboprop-powered aircraft like the Cessna Caravan and Quest Kodiak are used as bush airplanes.

A military transport aircraft, over 2,500 Lockheed C-130 Hercules have been built.

Turboprop engines are generally used on small subsonic aircraft, but the Tupolev Tu-114 can reach 470 kn (870 km/h, 541 mph). Large military aircraft, like the Tupolev Tu-95, and civil aircraft, such as the Lockheed L-188 Electra, were also turboprop powered. The Airbus A400M is powered by four Europrop TP400 engines, which are the second most powerful turboprop engines ever produced, after the 11 MW (15,000 hp) Kuznetsov NK-12.

The Beech King Air and Super King Air are the most-delivered turboprop business aircraft, with a combined 7,300 examples.

In 2017, the most widespread turboprop airliners in service were the ATR 42/72 (950 aircraft), Bombardier Q400 (506), De Havilland Canada Dash 8-100/200/300 (374), Beechcraft 1900 (328), de Havilland Canada DHC-6 Twin Otter (270), Saab 340 (225). Less widespread and older airliners include the BAe Jetstream 31, Embraer EMB 120 Brasilia, Fairchild Swearingen Metroliner, Dornier 328, Saab 2000, Xian MA60, MA600 and MA700, Fokker 27 and 50.

Turboprop business aircraft include the Piper Meridian, Socata TBM, Pilatus PC-12, Piaggio P.180 Avanti, Beechcraft King Air and Super King Air. In April 2017, there were 14,311 business turboprops in the worldwide fleet.

## Reliability

Between 2012 and 2016, the ATSB observed 417 events with turboprop aircraft, 83 per year, over 1.4 million flight hours: 2.2 per 10,000 hours. Three were "high risk" involving engine malfunction and unplanned landing in singleengine Cessna 208 Caravans, four "medium risk" and 96% "low risk". Two occurrences resulted in minor injuries due to engine malfunction and terrain collision in agricultural aircraft and five accidents involved aerial work: four in agriculture and one in an air ambulance.

## Turboshaft

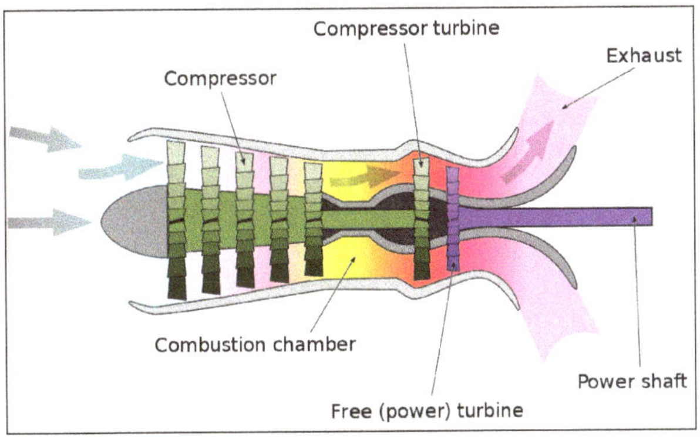


The operation of a simplified turboshaft engine.
The compressor spool is shown in green and the free / power spool is in purple.

A turboshaft engine is a form of gas turbine that is optimized to produce shaftpower rather than jet thrust.

In concept, turboshaft engines are very similar to turbojets, with additional turbine expansion to extract heat energy from the exhaust and convert it into output shaft power. They are even more similar to turboprops, with only minor differences, and a single engine is often sold in both forms.

Turboshaft engines are commonly used in applications that require a sustained high power output, high reliability, small size, and light weight. These include helicopters, auxiliary power units, boats and ships, tanks, hovercraft, and stationary equipment.

A turboshaft engine may be made up of two major parts assemblies: the 'gas generator' and the 'power section'. The gas generator consists of the compressor, combustion chambers with ignitors and fuel nozzles, and one or more stages of turbine. The power section consists of additional stages of turbines, a gear reduction system, and the shaft output. The gas generator creates the hot expanding gases to drive the power section. Depending on the design, the engine accessories may be driven either by the gas generator or by the power section.

In most designs, the gas generator and power section are mechanically separate so they can each rotate at different speeds appropriate for the conditions, referred to as a 'free power turbine'. A free power turbine can be an extremely useful design feature for vehicles, as it allows the design to forgo the weight and cost of complex multiple-ratio transmissions and clutches.

The general layout of a turboshaft is similar to that of a turboprop. The main difference is a turboprop is structurally designed to support the loads created by a rotating propeller, as the propeller is not attached to anything but the engine itself. In contrast, turboshaft engines usually drive a transmission which is not structurally attached to the engine. The transmission is attached to the vehicle structure and supports the loads created instead of the engine. In practice, though, many of the same engines are built in both turboprop and turboshaft versions, with only minor differences.

An unusual example of the turboshaft principle is the Pratt & Whitney F135-PW-600 turbofan engine for the STOVL F-35B – in conventional mode it operates as a turbofan, but when powering the LiftFan, it switches partially to turboshaft mode to send 29,000 horsepower forward through a shaft and partially to turbofan mode to continue to send thrust to the main engine's fan and rear nozzle.

Large helicopters use two or three turboshaft engines for redundancy. The Mil Mi-26 uses two Lotarev D-136 at 11,400 hp each, while the Sikorsky CH-53E Super Stallion uses three General Electric T64 at 4,380 hp each.

Early turboshaft engines were adaptations of turboprop engines, delivering power through a shaft driven directly from the gas generator shafts, via a reduction gearbox. Examples of direct-drive turboshafts include marinised or industrial Rolls-Royce Dart engines.

Austin 250hp gas turbine, sectioned.

The first examples of a gas turbine engine design ever considered for armoured fighting vehicles, the BMW 003-based GT 101, were tested in Nazi Germany's Panther tanks in mid-1944. The first true turboshaft engine for helicopters was built by the French engine firm Turbomeca, led by the founder, Joseph Szydlowski. In 1948, they built the first French-designed turbine engine, the 100-shp 782. Originally conceived as an auxiliary power unit, it was soon adapted to aircraft propulsion, and found a niche as a powerplant for turboshaft-driven helicopters in the 1950s. In 1950, this work was used to develop the larger 280-shp Artouste, which was widely used on the Aérospatiale Alouette II and other helicopters. This was following the experimental installation of a Boeing T50 turboshaft in an example of the Kaman K-225 synchropter on December 11, 1951, as the world's first-ever turboshaft-powered helicopter of any type to fly. The T-80 tank, which entered service with the Soviet Army in 1976, was the first tank using a gas turbine as the main engine. Since 1980 the US Army has operated the M1 Abrams tank. Designed and built in the late 1970s, this tank makes use of a gas turbine engine compared to most tanks using reciprocating diesels. The engine has considerably fewer parts, mechanically is very reliable, produces reduced exterior noise, and runs on virtually any fuel: petrol (gasoline), diesel fuel, aviation fuels. Similarly, the Swedish Stridsvagn 103 was the first tank to utilize a gas turbine, in this case as a secondary, high-horsepower "sprint" engine to augment the primary piston engine's performance.

# OTHER ENGINE TYPES

## Wankel Engines

Wankel RC2-60 Aeronautical Rotary Engine.

In principle, Wankel engines are ideal for light aircraft, being light, compact, almost vibrationless, and with a high power-to-weight ratio. Further aviation benefits of a Wankel engine include:

- Rotors cannot seize, since rotor casings expand greater than rotors.
- The engine is less prone to the serious condition known as "engine-knock", which can destroy plane's piston engines in mid-flight.
- The engine is not susceptible to "shock-cooling" during descent.
- The engine does not require an enriched mixture for cooling at high power.

- Having no reciprocating parts, there is less vulnerability to damage when the engine revolves at a higher rate than the designed maximum. The limit to the revolutions is the strength of the main bearings.

Unlike cars and motorcycles, a Wankel aero-engine will be sufficiently warm before full power is asked of it because of the time taken for pre-flight checks. A Wankel aero-engine spends most of its operational time at high power outputs, with little idling. This makes ideal the use of peripheral ports. An advantage is that modular engines with more than two rotors are feasible. If icing of any intake tracts is an issue, there is plenty of waste engine heat available to prevent icing.

The first Wankel rotary-engine aircraft was in the late 1960s being the experimental Lockheed Q-Star civilian version of the United States Army's reconnaissance QT-2, essentially a powered Schweizer sailplane. The plane was powered by a 185 hp (138 kW) Curtiss-Wright RC2-60 Wankel rotary engine. The same engine model was also used in a Cessna Cardinal and a helicopter, as well as other airplanes. In Germany in the mid-1970s, a pusher ducted fan airplane powered by a modified NSU multi-rotor Wankel engine was developed in both civilian and military versions, Fanliner and Fantrainer.

At roughly the same time as the first experiments with full-scale aircraft powered with Wankel engines, model aircraft-sized versions were pioneered by a combine of the well-known Japanese O.S. Engines firm and the then-extant German Graupner aeromodeling products firm, under license from NSU/Auto-Union. By 1968 the first prototype air-cooled, single-rotor glow plug-ignition, methanol-fueled 4.9 $cm^3$ displacement OS/Graupner model Wankel engine was running, and was produced in at least two differing versions from 1970 to the present day, solely by the O.S. firm after Graupner's demise in 2012.

Aircraft Wankel engines are increasingly being found in roles where the compact size, high power-to-weight ratio and quiet operation are important, notably in drones and unmanned aerial vehicles. Many companies and hobbyists adapt Mazda rotary engines, taken from cars, to aircraft use. Others, including Wankel GmbH itself, manufacture Wankel rotary engines dedicated for that purpose. One such use is the "Rotapower" engines in the Moller Skycar M400. Another example of purpose-built aircraft rotaries are Austro Engine's 55 hp (41 kW) AE50R (certified) and 75 hp (56 kW) AE75R (under development) both appr. 2 hp/kg.

Wankel engines are also becoming increasingly popular in homebuilt experimental aircraft, such as the ARV Super2 which can be re-engined with the British MidWest AE series aero-engine. Most are Mazda 12A and 13B automobile engines, converted to aviation use. This is a very cost-effective alternative to certified aircraft engines, providing engines ranging from 100 to 300 horsepower (220 kW) at a fraction of the cost of traditional piston engines. These conversions were initially in the early 1970s. With a number of these engines mounted on aircraft, as of 10 December 2006 the National Transportation Safety Board has only seven reports of incidents involving aircraft with Mazda engines, and none of these were a failure due to design or manufacturing flaws.

Mazda rotaries have worked well when converted for use in homebuilt aircraft. However, the real challenge in aviation is to produce FAA-certified alternatives to the standard reciprocating engines that power most small general aviation aircraft. Mistral Engines, based in Switzerland, developed purpose-built rotaries for factory and retrofit installations on certified production aircraft. The G-190 and G-230-TS rotary engines were already flying in the experimental market, and Mistral Engines hoped for FAA and JAA certification by 2011. As of June 2010, G-300 rotary engine development ceased, with the company citing cash flow problems.

Mistral claims to have overcome the challenges of fuel consumption inherent in the rotary, at least to the extent that the engines are demonstrating specific fuel consumption within a few points of reciprocating engines of similar displacement. While fuel burn is still marginally higher than traditional engines, it is outweighed by other beneficial factors.

At the price of increased complication for a high pressure diesel type injection system, fuel consumption in the same range as small pre-chamber automotive and industrial diesels has been demonstrated with Curtiss-Wright's stratified charge multi-fuel engines, while preserving Wankel rotary advantages Unlike a piston and overhead valve engine, there are no valves which can float at higher rpm causing loss of performance. The Wankel is a more effective design at high revolutions with no reciprocating parts, far fewer moving parts and no cylinder head.

The French company Citroën had developed Wankel powered RE-2 [fr] helicopter in the 1970s.

Since Wankel engines operate at a relatively high rotational speed, at 6,000 rpm of output shaft, the Rotor makes only 2,000 turns. With relatively low torque, propeller driven aircraft must use a propeller speed reduction unit to maintain propellers within the designed speed range. Experimental aircraft with Wankel engines use propeller speed reduction units, for instance the MidWest twin-rotor engine has a 2.95:1 reduction gearbox. The rotational shaft speed of a Wankel engine is high compared to reciprocating piston designs. Only the eccentric shaft spins fast, while the rotors turn at exactly one-third of the shaft speed. If the shaft is spinning at 7,500 rpm, the rotors are turning at a much slower 2,500 rpm.

Pratt & Whitney Rocketdyne has been commissioned by DARPA to develop a diesel Wankel engine for use in a prototype VTOL flying car called the "Transformer". The engine, based on an earlier unmanned aerial vehicle Wankel diesel concept called "Endurocore".

The sailplane manufacturer Schleicher uses Wankel engines in its self-launching models ASK-21 Mi, ASH-26E, ASH-25 M/Mi, ASH-30 Mi, ASH-31 Mi, ASW-22 BLE, and ASG-32 Mi.

In 2013 e-Go aeroplanes, based in Cambridge, United Kingdom, announced that its new single-seater canard aircraft, the winner of a design competition to meet the new UK single-seat deregulated category, will be powered by a Wankel engine from Rotron

Power, a specialist manufacturer of advanced rotary engines for unmanned aeronautical vehicle (UAV) applications. The first sale was 2016. The aircraft is expected to deliver 100 knots (190 km/h; 120 mph) cruise speed from a 30 hp (22 kW) Wankel engine, with a fuel economy of 75 $mpg_{imp}$ (3.8 L/100 km; 62 $mpg_{US}$) using standard motor gasoline (MOGAS), developing 22 kW (30 hp).

The DA36 E-Star, an aircraft designed by Siemens, Diamond Aircraft and EADS, employs a series hybrid powertrain with the propeller being turned by a Siemens 70 kW (94 hp) electric motor. The aim is to reduce fuel consumption and emissions by up to 25 percent. An onboard 40 hp (30 kW) Austro Engines Wankel rotary engine and generator provides the electricity. A propeller speed reduction unit is eliminated. The electric motor uses electricity stored in batteries, with the generator engine off, to take off and climb reducing sound emissions. The series-hybrid powertrain using the Wankel engine reduces the weight of the plane by 100 kg compared with its predecessor. The DA36 E-Star first flew in June 2013, making this the first ever flight of a series-hybrid powertrain. Diamond Aircraft state that the technology using Wankel engines is scalable to a 100-seat aircraft.

## Aircraft Diesel Engine

Thielert Centurion aircraft diesel engine.

The aircraft diesel engine or aero diesel has not been widely used as an aircraft engine. Diesel engines were used in airships and were tried in aircraft in the late 1920s and 1930s, but never widely used. Their main advantages are their excellent specific fuel consumption, the reduced flammability and somewhat higher density of their fuel, but these have been outweighed by a combination of inherent disadvantages compared to gasoline-fueled or turboprop engines. The ever-rising cost of avgas and doubts about its future availability have spurred a resurgence in aircraft diesel engine production in the early 2010s.

Using diesel engines in aircraft is additionally advantageous from the standpoint of environmental protection as well as the protection of human health, since the tetraethyllead antiknock ingredient of avgas has long been known to be highly toxicas well as polluting.

## Experimental Engines

A number of other manufacturers are currently developing experimental diesel engines, many using aircraft-specific designs rather than adapted automotive engines. Many are using two-stroke designs, with some opposed-piston layouts directly inspired by the original Junkers design.

Diesel Air Limited, Wilksch and Zoche have all had considerable problems bringing their prototype designs into production, with delays running into several years. The Diesel Air Limited-powered airship is no longer registered by the Civil Aviation Authority in the UK.

## Two-stroke

Wilksch Airmotive, a British company, is developing a 120 hp (89 kW) three-cylinder two-stroke diesel (WAM-120) and is working on a four-cylinder 160 hp (120 kW) design (WAM-160). In 2007 Wilksch claimed that they had completed multiple tests on the WAM-100 LSA in accordance with ASTM F 2538 – the WAM-100 LSA is a de-rated WAM-120. Wilksch originally showed a two-cylinder prototype alongside the three- and four-cylinder models. In April 2008 IndUS Aviation introduced the first diesel light-sport aircraft with a WAM 120 having flown 400 hours on a Thorp T211 in England for the past four years. By mid-2009, approximately 40 WAM-120 units had been sold, with around half currently flying. The British owner of a VANS RV-9A fitted with a WAM-120 reports getting 125 kn (232 km/h) TAS at 6,000 ft (1,800 m) on 15 litre/hr of Jet A-1 fuel. A Rutan LongEz canard-pusher (G-LEZE) has also flown with the WAM-120 engine with test flights showing a TAS of 160 kn (300 km/h) at 11,000 ft (3,400 m) and 22 litre/hr. At economy cruise of 125 kn (232 km/h) at 2,000 ft (610 m) the fuel consumption is 12 litre/hr, giving a range of 1,890 nmi (3,500 km).

An American company, is currently developing three V-4 designs of 160, 180 and 200 hp (150 kW), the latter two versions being turbocharged. Using a ported two-stroke design, they have also flown a prototype engine in a pusher configuration. Velocity aircraft are claiming delivery of non-certified engines since 2005 and hope to achieve certification early in 2011. DeltaHawk engines have a dry oil sump, so they can run in any orientation: upright, inverted or vertical shaft by changing the location of the oil scavenge port. They can also run counter-rotation for installation in twins to eliminate the critical engine issue. A water-cooled DeltaHawk engine has been successfully fitted to a Rotorway helicopter, weighing the same as an air-cooled petrol engine of similar power and being capable of maintaining that power to 17,000 feet. GAP Diesel Engine is a NASA development.

With the branding Zoche aero-diesel, the company “Michael Zoche Antriebstechnik” in Munich/Germany has produced a prototype range of three radial air-cooled two-stroke diesel aero-engines, comprising a V-twin, a single-row cross-4 and a double-row cross-8. A Zoche engine has run successfully in wind tunnel tests. Zoche seem barely closer to production than they were a decade ago.

Andy Higgs' Advanced Component Engineering designed a step piston 1,000 hp (750 kW) V12 weighing 665 lb/302 kg with the reduction gearbox to replace low-end, 580 to 1,200 hp PT6s like in the Cessna Caravan; a 350 hp (260 kW), 302 pounds/137 kg four-cylinder with a gearbox to reduce the prop. rpm to 2300 from 5300; and a 1.5-liter V4 weighing 103 pounds/47 kg and producing 120 hp (89 kW). The 1,000 in$^3$ (16 l) v12 can power generators, tanks, boats or blimps and v4 and v8 versions can be derived.

## Opposed-piston Engines

The Bourke engine, designed by Russell Bourke, of Petaluma, CA, is an opposed rigidly connected twin cylinder design using the detonation principle.

Diesel Air Limited is a British company developing a 100 hp (75 kW) twin-cylinder (therefore four-piston), two-stroke opposed-piston engine inspired by the original Junkers design. Their engine has flown in test aircraft and airship installations. Unlike the Junkers, it is made for horizontal installation with a central output shaft for the geared cranks, the overall installed shape thereby approximately resembling a four-stroke flat-four engine.

Powerplant Developments, a British company, is developing a 100 and 120 hp (75 and 89 kW) opposed-piston engine called the Gemini 100/120 that resembles the Diesel Air Limited engine and uses the Junkers twin-crank principle, again for horizontal installation with a central output shaft for the geared cranks. However, the Gemini 100 is an engine. Like Diesel Air Limited, Powerplant Developments claim to be using Weslake Air Services for production. They have recently announced that Tecnam will test a prototype with the Gemini engine.

Superior Air Parts' subsidiary Gemini Diesel develops three cylinder two-stroke designs with six opposed pistons, a 100 hp / 75 kW weighing 159.5 lb / 72.5 kg, and a turbocharged 125 hp / 118 kW weighing 175 lb / 72.5 kg, both measuring 23" wide × 16" high × 23" long (58 × 40 cm × 58 cm) and reaching.38 and.378 lb/hp/h (231 and 230 g/kW/h) BSFC, respectively; while larger three cylinder, six pistons engine would produce 180-200 hp / 134-149 kW whilst weighing 276 lb / 125 kg in 29" W × 16" H × 29" L (73 × 41.5 × 72.5 cm) and 300-360 HP / 224-268 KW turbocharged whilst weighing 386 lb / 175 kg within 29" W × 19" H × 37" L (73 × 47.5 × 95 cm), while five cylinders, 10 pistons would produce 450 hp / 336 kW whilst weighing 474 lb / 215 kg within 29" W × 22" H × 43 " L (73 × 55 × 110 cm) and six cylinder, 12 pistons engines would attain 550 hp / 410 kW whilst weighing 551 pounds / 250 kg in 29" W × 22" H × 48" L (73 × 55 × 122 cm), burning.386 to.360 lb/hp/h (235 to 219 g/kW/h) The 100 hp version will cost less than $25,000.

Weslake Engine, another UK based company, displayed its A80 lightweight diesel engine at Friedrichshafen Aero 2015.

## Four-stroke

Wisconsin-based Engineered Propulsion Systems develops its Graflight liquid-cooled V-8 engine with steel pistons and compacted graphite iron crankcase for better strength and durability than aluminum at similar weight, increasing time between overhauls to 3,000 h. It is managed by a Bosch ECU and consumes Jet A, JP-8 or straight diesel for general aviation aircraft and small helicopters, military drones, small boats or troop carriers, and its low vibration allows the use of composite or aluminum airscrews. At 262 hp (195 kW), 75% of the 350 hp (261 kW) maximum power, it consumes 77 lb/h (35 kg/h), in comparison to the Continental TSIO-550-E, which burns 110 lb/h (50 kg/h).

## Automotive Derived

Raptor Turbo Diesel LLC, an American company, is currently developing the Raptor 105 diesel engine. It is a four-stroke inline turbo charged engine. Formerly Vulcan Aircraft Engines.

ECO Motors developed the EM 80 and EM 100 hp (75 kW) 4 stroke 4 cylinder diesel with FADEC based on a car engine for 98 kg (216 lb) dry but disappeared since 2008.

The FlyEco diesel is a three-cylinder, 0.8L engine producing 80 HP/58,8 kW till 3,800 rpm and reduced by 1:1.50-1.79, derived from the Smart Car. It powers the Siemens-FlyEco Magnus eFusion hybrid electric aircraft.

## Teos/Austro Engine AE440

The AE440 presented by Mecachrome.

Within the Green Rotorcraft European Clean Sky Joint Technology Initiative environmental research program started in 2011, an Airbus Helicopters H120 Colibri technology demonstrator equipped with a HIPE AE440 high-compression diesel engine, running on jet fuel, first flew on 6 November 2015. The powerplant is a liquid-cooled, dry sump lubricated 4.6-liter 90° V8 engine with 1800 bar common rail direct injection, fully machined aluminum blocks, titanium connecting rods, steel pistons and liners, one turbocharger per cylinder bank.

With an air/air intercooler, it weighs 197 kg (dry) without gearbox and the 330 kW (440 hp) installed powerpack weighs 249 kg. Its brake specific fuel consumption is 200 g/kW.h. It is manufactured by Teos Powertrain Engineering, a joint venture between Mecachrome and D2T (IFPEN group) for the mechanical design, engine main parts manufacturing, assembly and testing and Austro Engine for the dual channel FADEC and harness, fuel system, airworthiness.

## Precooled Jet Engines

The precooled jet engine is a concept that enables jet engines with turbomachinery, as opposed to ramjets, to be used at high speeds. Precooling restores some or all of the performance degradation of the engine compressor (by preventing rotating stall/choking/reduced flow), as well as that of the complete gas generator (by maintaining a significant combustor temperature rise within a fixed turbine temperature limit), which would otherwise prevent flight with high ram temperatures.

For higher flight speeds precooling may feature a cryogenic fuel-cooled heat exchanger before the air enters the compressor. After gaining heat and vapourising in the heat exchanger, the fuel (e.g. $H_2$) burns in the combustor. Precooling using a heat exchanger has not been used in flight, but is predicted to have significantly high thrust and efficiency at speeds up to Mach 5.5. Precooled jet engine cycles were analyzed by Robert P. Carmichael in 1955. Pre-cooled engines avoid the need for an air condenser because, unlike liquid air cycle engines (LACE), pre-cooled engines cool the air without liquefying it.

For lower flight speeds precooling can be done with mass injection, known as WIPCC (water injection precompressor cooling). This method has been used for short duration (due to limited coolant capacity) increases to an aircraft's normal maximum speed. "Operation Skyburner", which gained a world speed record with a McDonnell Douglas F-4 Phantom II, and the Mikoyan Ye-266 (Mig 25). Both used a water/alcohol spray to cool the air ahead of the compressor.

Precooling (as well as combustion chamber water injection) is used at the lowest flight speeds, i.e. during takeoff, to increase thrust at high ambient temperatures.

### Advantages and Disadvantages of using Precooler Heat Exchangers

One main advantage of pre-cooling is (as predicted by the ideal gas law) for a given overall pressure ratio, there is a significant reduction in compressor delivery temperature (T3), which delays reaching the T3 limit to a higher Mach number. Consequently, sea-level conditions (corrected flow) can be maintained after the pre-cooler over a very wide range of flight speeds, thus maximizing net thrust even at high speeds. The compressor and ducting after the inlet is subject to much lower and more consistent temperatures, and hence may be made of light alloys. This reduces the weight of the engine, which further improves the thrust/weight ratio.

Hydrogen is a suitable fuel because it is liquid at deeply cryogenic temperatures, and over its useful range has a very high total specific heat capacity, including the latent heat of vapourisation, higher than water.

However, the low density of liquid hydrogen has negative effects on the rest of the vehicle, and the vehicle physically becomes very large, although the weight on the undercarriage and wing loading may remain low.

Hydrogen causes structural weakening in many materials, known as hydrogen embrittlement.

The weight of the precooler adds to the weight of the engine, thereby reducing its thrust to weight ratio.

Passing the intake air through the precooler adds to the inlet drag, thereby reducing the engine net thrust, and so reducing the thrust to weight ratio.

Depending on the amount of cooling required, despite its high thermal capacity, more hydrogen may be needed to cool the air than can be burnt with the cooled air. In some cases, part of the excess hydrogen can be burnt in a ramjet with uncooled air to reduce this inefficiency.

Unlike a LACE engine, a precooled engine doesn't need to liquefy the oxygen, so the amount of cooling is reduced as there is no need to cover of fusion of the oxygen and a smaller total temperature drop is required. This in turn reduces the amount of hydrogen used as a heat-sink, but unable to be burnt. In addition a condenser isn't required, giving a weight saving.

## References


- Piston-Engine: skybrary.aero, Retrieved 16 April, 2020
- Erjavec, Jack (2010). Automotive Technology: A Systems Approach. Clifton Park, NY USA: Delmar, Cengage Learning. pp. 226–227. ISBN 978-1428311497. LCCN 2008934340. Retrieved 2014-02-09
- Types-of-aircraft-piston-engines: blog.covingtonaircraft.com, Retrieved 27 March, 2020
- Sonntag, Richard E.; Borgnakke, Claus (2006). Introduction to engineering thermodynamics (Second ed.). John Wiley. ISBN 9780471737599
- Pulse-jet-engine: appropedia.org, Retrieved 13 July, 2020
- Frank Northen Magill, ed. (1993). Magill's Survey of Science: Applied science series, Volume 3. Salem Press. p. 1431. ISBN 9780893567088. "Most tactical military aircraft are powered by low-bypass turbofan engines"
- El-Sayed, Ahmed F. (July 6, 2017). Aircraft propulsion and gas turbine engines (2nd ed.). CRC Press. Table 6.11. ISBN 9781466595187. OCLC 986784025

# 5

# Aircraft Systems

Various systems and its instruments are used in aircrafts such as pilot-static systems and gyroscopic systems. This chapter closely examines different aircraft systems to provide an easy understanding of the subject.

## PITOT-STATIC SYSTEMS

Pitot-static systems measure and compare air pressures from the Pitot tube and static port. These measurements are typically used to display indications for the Airspeed Indicator (ASI), Altimeter, and Vertical Speed Indicators (VSI). May be heated to provide protection from ice buildup.

Consists of two components:

- Pitot Tube,
- Static Port.

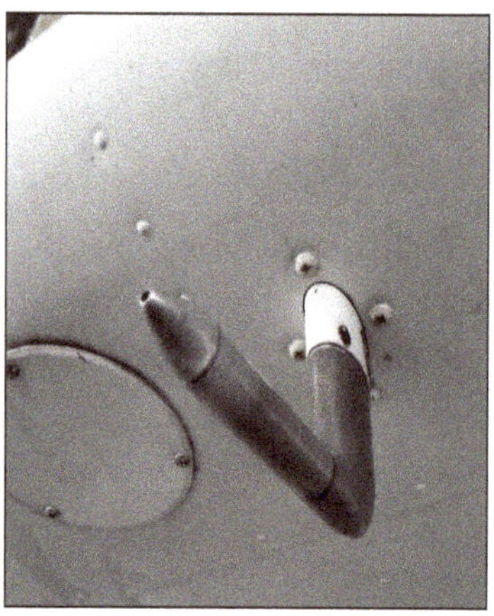

Pitot Tube.

Static Port.

Pitot pressure, or impact air pressure, is sensed through an open-end tube pointed directly into the relative wind flowing around the aircraft. The pitot tube connects to the ASI or an air data computer depending on your aircraft's configuration.

## Static Pressure

Static pressure is also used by the ASI as well as the other pitot static instruments for determining altitude and vertical speed. Static pressure may be sensed at one or more locations on an aircraft. Some may be flush mounted on the fuselage or integrated into the electrically heated pitot tube. These ports are in locations proven by flight tests to be in undisturbed air, and they may be paired, one on either side of the aircraft. This dual location prevents lateral movement of the aircraft from giving erroneous static pressure indications. The areas around the static ports may be heated with electric heater elements to prevent ice forming over the port and blocking the entry of the static air.

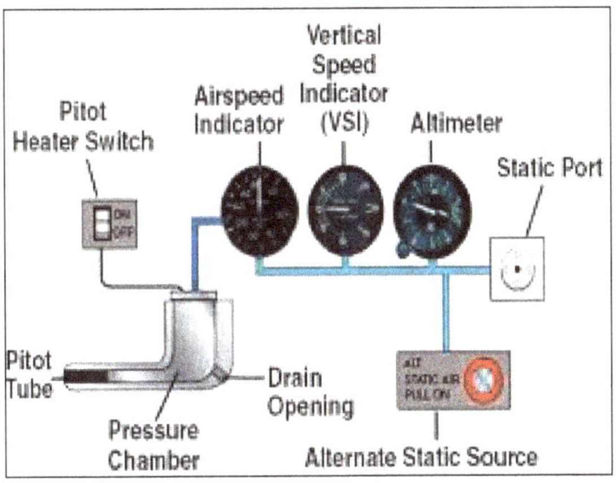

A typical electrically heated pitot-static head.

Three basic pressure-operated instruments are found in aircraft instrument panels flown under IFR. These are the ASI, sensitive altimeter, and vertical speed indicator (VSI). All three instruments receive static air pressure for operation with only the ASI receiving both pitot and static pressure.

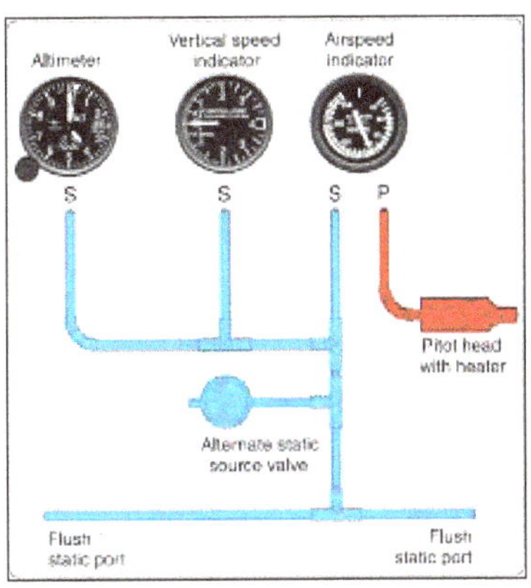

A typical pitot-static system.

## Blockage of the Pitot-Static System

Errors in the ASI and VSI almost always indicate a blockage of the pitot tube, the static port(s), or both. Moisture (including ice), dirt, or even insects can cause a blockage in both systems. During preflight, it is very important to make sure the pitot tube cover is removed and that static port openings are checked for blockage and damage.

## Blocked Pitot System

If the pitot tube drain hole becomes obstructed, the pitot system can become partially or completely blocked. When dynamic pressure cannot enter the pitot tube opening, the ASI no longer operates. If the drain hole is open, static pressure equalizes on both sides of the diaphram in the ASI and the indicated airspeed slowly drops to zero. If the pitot tube ram pressure hole and drain hole become obstructed, the ASI operates like an altimeter as the aircraft climbs and descends.

## Blocked Static System

When a static system becomes blocked but the pitot tube remains clear the ASI continues to operate but is inaccurate. When the aircraft is operated above the altitude where the static ports became blocked the airspeed indicates lower than the actual airspeed because the trapped static pressure is higher than normal for that altitude. The opposite holds true for operations at lower altitudes; a faster than actual airspeed is displayed due to the relatively low static pressure trapped in the system.

A blockage of the static system can also affect the altimeter and VSI. Trapped static pressure causes the altimeter to freeze at the altitude where the blockage occurred. In the case of the VSI, a blocked static system produces a continuous zero indication. An alternate static source is provided in some aircraft to provide static pressure should the primary static source become blocked. The alternate static source is normally found inside of the flight deck. Due to the venturi effect of the air flowing around the fuselage, the air pressure inside the flight deck is lower than the exterior pressure.

When the alternate static source pressure is used, the following instrument indications are observed:

- The altimeter indicates a slightly higher altitude than actual.
- The ASI indicates an air speed greater than the actual airspeed.
- The VSI shows a momentary climb and then stabilizes if the altitude is held constant.

## Effects of Flight Conditions

The static ports are located in a position where the air at their surface is as undisturbed as possible. But under some flight conditions, particularly at a high angle of attack with the landing gear and flaps down, the air around the static port may be disturbed to the extent that it can cause an error in the indication of the altimeter and ASI. Because of the importance of accuracy in these instruments, part of the certification tests for an aircraft is a check of position error in the static system.

## Altimeter

Altimeter is an instrument that measures the altitude of the land surface or any object such as an airplane. The two main types are the pressure altimeter, or aneroid barometer, which approximates altitude above sea level by measuring atmospheric pressure, and the radio altimeter, which measures absolute altitude (distance above land or water) based on the time required for a radio wave signal to travel from an airplane, a weather balloon, or a spacecraft to the ground and back.

The pressure altimeter operates on the principle that average atmospheric pressure decreases linearly with altitude. A typical pressure altimeter is illustrated in the figure. The instrument is enclosed in a case that is connected to the outside of the aircraft by an air pressure inlet at the rear of the housing. Two or more aneroid capsules—i.e., thin corrugated metallic bellows from which air has been exhausted—are positioned near the inlet. These capsules expand when the outside air pressure falls (as in climbing) and contract when the outside air pressure rises (as in descending). By a mechanical arrangement of sector gears, pinion, backlash spring, and crankshaft, the expansion or contraction of the aneroid capsules is converted to the movement of pointers on a dial. The graduated scale dial is marked off in metres or feet, and a series of gear-driven pointers similar to the hands of a clock may be used to indicate the altitude in units of hundreds, thousands, or tens of thousands. The barometric scale dial records the air pressure in millibars (mb). Because atmospheric pressure is measured relative to sea level, a pressure altimeter must be adjusted with a barosetting knob in order to compensate for small variations in barometric pressure caused by changes in local weather.

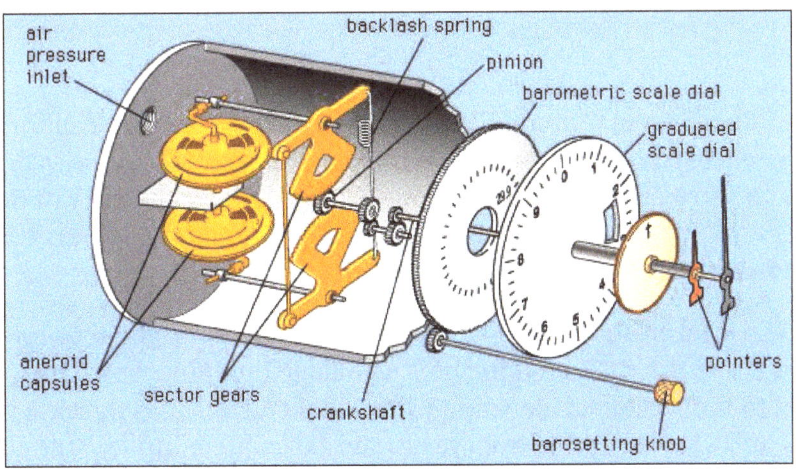


Pressure altimeter.

The radio altimeter measures the distance of an aircraft above the ground rather than above sea level. The altitude is equal to one-half the time that it takes a pulse of radio energy to travel from the aircraft to the ground and back multiplied by the speed of the pulse (equivalent to the speed of light). The measured altitude is displayed on a video screen. Radio altimeters are used in automatic navigation and blind-landing systems.

## Altimeter Purpose

Altimeters exist as a means of maintaining altitude. Air Traffic Control issues altimeter (pressure) settings to pilots of a local area to ensure everyone is using the same reference datum. This is especially important during Instrument Flight Rules where separation responsibilities fall upon ATC. Unless under control of ATC, altitudes are not assigned to pilot's however, to ensure safety and instrumentation accuracy, local altimeter settings are still very important.

- Instances where VFR pilots need to be accurate with their altitude include flying VFR cruising altitudes, flying traffic patterns, and maintaining minimum heights above protected areas.

## Altimeter Function

The altimeter functions like a barometer, detecting pressure differences from a standard (29.92 in-Hg, aka, absolute pressure) using the static system. The altimeter can be calibrated for non-standard pressure by changing the pressure setting in the Kollsman window.

Rotating the knob changes both the barometric scale and the altimeter pointers in such a way that a change in the barometric scale of 1" Hg changes the pointer indication by 1,000' which indicates the standard pressure lapse rate. The scale is set based on the local AWOS/ASOS/ATIS reading.

## Function

As an aircraft increases in altitude the aneroid capsules which are calibrated to 29.92 expand due to decreasing ambient air pressure which moves the needles through gears and linkages to show an increase in altitude. Conversely, as the aircraft descends, increasing pressure on the capsules causes them to contract while gears and linkages display a decrease in altitude.

When set to the local altimeter setting, the altimeter shows height above sea level (true altitude), but not above such land features as hills, mountains, or valleys (absolute altitude). If set to the local field elevation prior to takeoff, such as the case with gliders, the altimeter will show height above the ground (absolute altitude), but not above sea level (sea level).

## Types of Altimeters

Different altimeters measure altitude through various methods such as radio, radar, laser, or capacitive technology.

### Drum Type

Instrument Flying Handbook, Drum-Type Altimeter.

- These instruments have only one pointer that makes one revolution for every 1,000'.
- Each number represents 100' and each mark represents 20'.
- A drum, marked in thousands of feet, is geared to the mechanism that drives the pointer.
- To read this type of altimeter, first look at the drum to get the thousands of feet, and then at the pointer to get the feet and hundreds of feet.

### Sensitive Altimeter

Instrument Flying Handbook, Three Pointer Altimeter.

- One with an adjustable barometric scale allowing the pilot to set the reference pressure from which the altitude is measured.
- This scale is visible in a small window called the Kollsman window.
- A knob on the instrument adjusts the scale.
- The range of the scale is from 28.00" to 31.00" inches of mercury (Hg), or 948 to 1,050 millibars.

## Absolute Altimeter

- Also referred to as a radio altimeter, indicating the altitude over objects below based on radio signal pulses.

## Types of Altitude

Altitude is a measure of how far you are above a reference point.

## Absolute Altitude

The height of an aircraft above the terrain over which it is flying, usually expressed in AGL (above ground level) (HAA, HAT, TCH).

## Indicated Altitude

After setting the altimeter at the current altimeter setting, this is the uncorrected altitude read directly from that altimeter.

## True Altitude

Actual aircraft's height above sea level, usually expressed as MSL (mean sea level); all elevations on aeronautical charts are expressed in terms of true altitude.

## Pressure Altitude

After adjusting the altimeter's setting to 29.92 mb or 1,013.2 mb, this is the altimeter reading that corresponds to the altitude in the standard atmosphere where the pressure is the same as you are.

## LAGS: (Inversely Proportional)

- If the given altimeter setting is less than 29.92, then you ADD to your altitude.
- If the given altimeter setting is greater than 29.92, then you SUBTRACT from your altitude.

## Density Altitude

- The pressure altitude corrected for temperature variations.
- This is most important to takeoff and climb performance of an aircraft.

## GPS Altitude

- GPS Altitude is a measure take from, and therefore in reference to, the GPS Satellite Constellation.
- Since pressure readings passed to pilots are taken on the ground, atmospherics will guarantee a disparity between GPS altitude and pressure altitude as altitude increases.
- Although GPS altitude can be highly accurate, it is usually not referenced while in flight.
- This ensures all aircraft are using the same reference, and any error from pressure settings is generally shared across all aircraft.
- Altimeter Construction.

Consists of evacuated, corrugated bronze aneroid capsules which connect to needles on the face of the instrument. A striped segment is visible below 10,000', while above, a mask begins to cover it until 15,000', where it is fully covered. Stripes start to cover at 10,000' and completely cover at 15,000'.

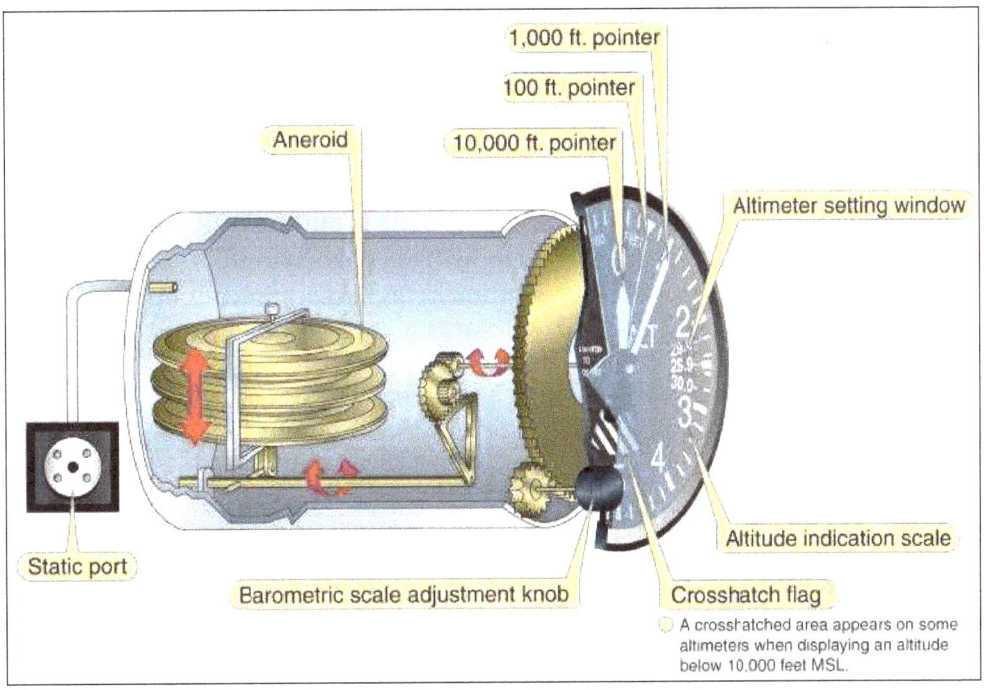


Instrument Flying Handbook, Sensitive Altimeter Components.

## Altimeter Errors

The accuracy of an aircraft's altimeter depends on a number of factors:

- Nonstandard atmospheric temperature.
- Nonstandard atmospheric pressure.
- Aircraft static pressure systems (position error).
- Instrument error.

## Non-standard Atmospheric Temperature

The crucial values to consider are standard temperature versus the ambient (at altitude) temperature:

- It is this "difference" that causes the error in indicated altitude.

  When the column of air is warmer than standard, you are higher than your altimeter indicates.
- Subsequently, when the column of air is colder than standard, you are lower than indicated.
- It is the magnitude of these differences that determine the magnitude of the error.
- When flying into a cooler air mass while maintaining a constant indicated altitude, you are losing true altitude. However, flying into a cooler air mass does not necessarily mean you will be lower than indicated if the difference is still on the plus side. For example, while flying at 10,000 feet (where STANDARD temperature is +5 degrees Celsius (C)), the outside air temperature cools from +5 degrees C to 0 degrees C, the temperature error will nevertheless cause the aircraft to be HIGHER than indicated. It is the extreme "cold" difference that normally would be of concern to the pilot. Also, when flying in cold conditions over mountainous terrain, the pilot should exercise caution in flight planning both in regard to route and altitude to ensure adequate en route and terminal area terrain clearance.

Non-standard temperatures can result in a change to effective vertical paths and actual descent rates while using aircraft Baro-VNAV equipment for vertical guidance on final approach segments. A higher than standard temperature will result in a steeper gradient and increased actual descent rate. Indications of these differences are often not directly related to vertical speed indications. Conversely, a lower than standard temperature will result in a shallower descent gradient and reduced actual descent rate. Pilots should consider potential consequences of these effects on approach minimums, power settings, sight picture, visual cues, etc., especially for high-altitude or terrain-challenged locations and during low-visibility conditions.

## ICAO Cold Temperature Error Table

The cold temperature induced altimeter error may be significant when considering obstacle clearances when temperatures are well below standard. Pilots may wish to increase their minimum terrain clearance altitudes with a corresponding increase in ceiling from the normal minimum when flying in extreme cold temperature conditions. Higher altitudes may need to be selected when flying at low terrain clearances. Most flight management systems (FMS) with air data computers implement a capability to compensate for cold temperature errors. Pilots flying with these systems should ensure they are aware of the conditions under which the system will automatically compensate. If compensation is applied by the FMS or manually, ATC must be informed that the aircraft is not flying the assigned altitude. Otherwise, vertical separation from other aircraft may be reduced, creating a potentially hazardous situation.

To use the table, find the reported temperature in the left column, and then read across the top row to the height above the airport/reporting station. Subtract the airport elevation from the altitude of the intended flight altitude (final approach fix [FAF]). The intersection of the column and row is the amount of possible error.

Example: The reported temperature is -10° Celsius and the FAF is 500' above the airport elevation.

The reported current altimeter setting may place the aircraft as much as 50' below the altitude indicated by the altimeter.

Pilots are responsible to compensate for cold temperature altimetry errors when operating into an airport with any published cold temperature restriction and a reported airport temperature at or below the published temperature restriction. Pilots must ensure compensating aircraft are correcting on the proper segment or segments of the approach. Manually correct if compensating aircraft system is inoperable. Pilots manually correcting, are responsible to calculate and apply a cold temperature altitude correction derived from to the affected approach segment or segments.

Pilots must advise the cold temperature altitude correction to Air Traffic Control (ATC). Pilots are not required to advise ATC of a cold temperature altitude correction inside of the final approach fix. When using the cold temperature error table, the altitude error is proportional to both the height above the reporting station elevation and the temperature at the reporting station. For IFR approach procedures, the reporting station elevation is assumed to be airport elevation. It is important to understand that corrections are based upon the temperature at the reporting station, not the temperature observed at the aircraft's current altitude and height above the reporting station and not the charted IFR altitude.

To see how corrections are applied, note the following examples:

- Airport Elevation: 496'.

- Airport Temperature: –50 °C.

A charted IFR approach to the airport provides the following data:

- Minimum Procedure Turn Altitude: 1,800'.
- Minimum FAF Crossing Altitude: 1,200'.
- Straight-in Minimum Descent Altitude: 800'.
- Circling MDA: 1,000'.

The Minimum Procedure Turn Altitude of 1,800' will be used as an example to demonstrate determination of the appropriate temperature correction. Typically, altitude values are rounded up to the nearest 100' level. The charted procedure turn altitude of 1,800' minus the airport elevation of 500' equals 1,300'. The altitude difference of 1,300' falls between the correction chart elevations of 1,000' and 1,500'. At the station temperature of -50 °C, the correction falls between 300' and 450'. Dividing the difference in compensation values by the difference in altitude above the airport gives the error value per foot. In this case, 150' divided by 500' = 0.33' for each additional foot of altitude above 1,000'.

This provides a correction of 300' for the first 1,000' and an additional value of 0.33 times 300', or 99', which is rounded to 100'. 300' + 100' = total temperature correction of 400'. For the given conditions, correcting the charted value of 1,800' above MSL (equal to a height above the reporting station of 1,300') requires the addition of 400'. Thus, when flying at an indicated altitude of 2,200', the aircraft is actually flying a true altitude of 1,800'. Minimum Procedure Turn Altitude 1,800' charted = 2,200' corrected.

Minimum FAF Crossing Altitude 1,200' charted = 1,500' corrected:

- Straight-in MDA 800' charted = 900' corrected.
- Circling MDA 1,000' charted = 1,200' corrected.

| Reported Temp C° | Height Above Airport in Feet | | | | | | | | | | | | | |
|---|---|---|---|---|---|---|---|---|---|---|---|---|---|---|
| | 200 | 300 | 400 | 500 | 600 | 700 | 800 | 900 | 1,000 | 1,500 | 2,000 | 3,000 | 4,000 | 5,000 |
| +10 | 10 | 10 | 10 | 10 | 20 | 20 | 20 | 20 | 20 | 30 | 40 | 60 | 80 | 90 |
| 0 | 20 | 20 | 30 | 30 | 40 | 40 | 50 | 50 | 60 | 90 | 120 | 170 | 230 | 280 |
| -10 | 20 | 30 | 40 | 50 | 60 | 70 | 80 | 90 | 100 | 150 | 200 | 290 | 390 | 490 |
| -20 | 30 | 50 | 60 | 70 | 90 | 100 | 120 | 130 | 140 | 210 | 280 | 420 | 570 | 710 |
| -30 | 40 | 60 | 80 | 100 | 120 | 130 | 150 | 170 | 190 | 280 | 380 | 570 | 760 | 950 |
| -40 | 50 | 80 | 100 | 120 | 150 | 170 | 190 | 220 | 240 | 360 | 480 | 720 | 970 | 1,210 |
| -50 | 60 | 90 | 120 | 150 | 180 | 210 | 240 | 270 | 300 | 450 | 590 | 890 | 1,190 | 1,500 |

Instrument Flying Handbook,ICAO Cold Temperature Error Table.

## Non-standard Atmospheric Pressure

Once in flight, it is very important to obtain frequently current altimeter settings en route. If you do not reset your altimeter when flying from an area of high pressure into an area of low pressure, your aircraft will be closer to the surface than your altimeter indicates. An inch error in the altimeter setting equals 1,000 feet of altitude.

TBL 7-2-3, derived from ICAO formulas, indicates how much error can exist when the temperature is extremely cold. To use the table, find the reported temperature in the left column, then read across the top row to locate the height above the airport/reporting station (i.e., subtract the airport/reporting elevation from the intended flight altitude). The intersection of the column and row is how much lower the aircraft may actually be as a result of the possible cold temperature induced error.

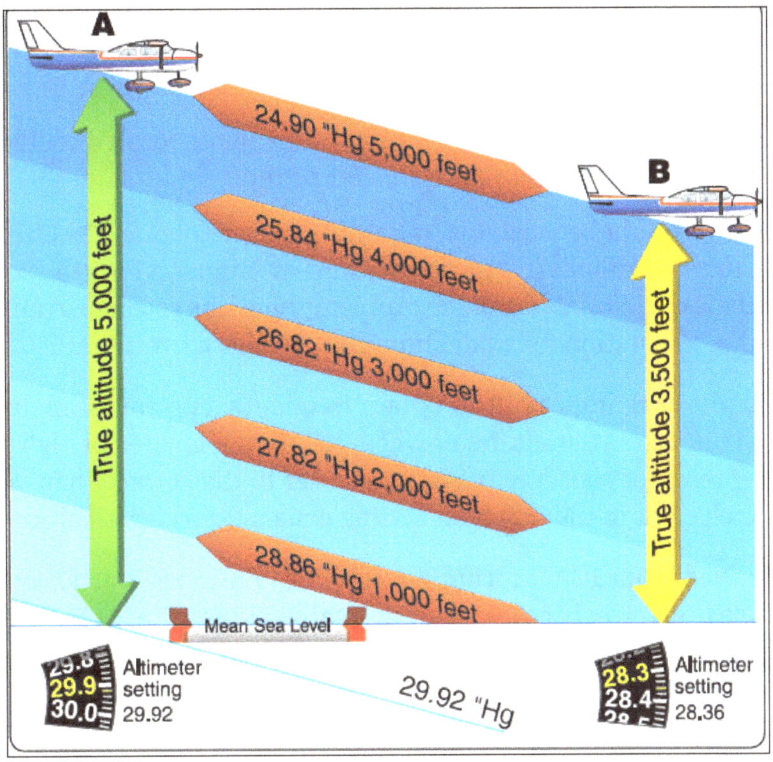


Instrument Flying Handbook, Effects of Nonstandard Pressure on an Altimeter of an Aircraft Flown into Air of Lower than Standard Pressure (Air is Less Dense).

## Position Error

Most aircraft do not have the capability of setting an altimeter above 31.00 in-Hg or below 28.00 in-Hg.

For aircraft with the capability of setting the current altimeter setting and operating into airports with the capability of measuring the current altimeter setting, no additional restrictions apply.

The possible result of the above example should be obvious, particularly if operating at the minimum altitude or when conducting an instrument approach. When operating in extreme cold temperatures, pilots may wish to compensate for the reduction in terrain clearance by adding a cold temperature correction. A sensitive altimeter is designed to indicate standard changes from standard conditions, but most flying involves errors caused by non-standard conditions and the pilot must be able to modify the indications to correct for these errors.

## Instrument Errors

Most pressure altimeters are subject to mechanical, elastic, temperature, and installation errors. Although manufacturing and installation specifications, as well as the periodic test and inspections required by regulations act to reduce these errors, any scale error may be observed in the following manner:

- Set the current reported altimeter setting on the altimeter setting scale.
- Altimeter should now read field elevation if you are located on the same reference level used to establish the altimeter setting.
- Note the variation between the known field elevation and the altimeter indication. If this variation is in the order of plus or minus 75 feet, the accuracy of the altimeter is questionable and the problem should be referred to an appropriately rated repair station for evaluation and possible correction.

Once in flight, it is very important to obtain frequently current altimeter settings en route. If you do not reset your altimeter when flying from an area of high pressure into an area of low pressure, your aircraft will be closer to the surface than your altimeter indicates. An inch error in the altimeter setting equals 1,000 feet of altitude.

TBL 7-2-3, derived from ICAO formulas, indicates how much error can exist when the temperature is extremely cold. To use the table, find the reported temperature in the left column, then read across the top row to locate the height above the airport/ reporting station (i.e., subtract the airport/ reporting elevation from the intended flight altitude). The intersection of the column and row is how much lower the aircraft may actually be as a result of the possible cold temperature induced error. Pilots are responsible to compensate for cold temperature altimetry errors when operating into an airport with any published cold temperature restriction and a reported airport temperature at or below the published temperature restriction. Pilots must ensure compensating aircraft are correcting on the proper segment or segments of the approach. Manually correct if compensating aircraft system is inoperable. Pilots manually correcting, are responsible to calculate and apply a cold temperature altitude correction derived from TBL 7-2-3 to the affected approach segment or segments. Pilots must advise the cold temperature altitude correction to Air Traffic Control (ATC). Pilots are not required to advise ATC of a cold temperature altitude correction inside of the final approach fix. The possible result of the above example should be obvious, particularly if operating at the minimum altitude or when conducting an instrument approach. When operating in extreme cold temperatures, pilots may wish to compensate for the reduction in terrain clearance by adding a cold temperature correction.

- Example: Temperature-10 degrees Celsius, and the aircraft altitude is 1,000 feet above the airport elevation. The chart shows that the reported current altimeter setting may place the aircraft as much as 100 feet below the altitude indicated by the altimeter

## Inherent Altimeter Errors

When the aircraft is flying in air that is warmer than standard, the air is less dense and the pressure levels are farther apart. When the aircraft is flying at an indicated altitude of 5,000', the pressure level for that altitude is higher than it would be in air at standard temperature, and the aircraft is higher than it would be if the air were cooler.

If the air is colder than standard, it is denser and the pressure levels are closer together. When the aircraft is flying at an indicated altitude of 5,000', its true altitude is lower than it would be if the air were warmer. Most pressure altimeters are subject to mechanical, elastic, temperature, and installation errors.

## Mechanical Errors

Table: Lowest usable flight level.

| Almeter Setting | Lowest Usable Flight Level |
|---|---|
| 29.92 or higher | 180 |
| 29.91 to 29.42 | 185 |
| 29.41 to 28.92 | 190 |
| 28.91 to 28.42 | 195 |
| 28.41 to 27.91 | 200 |

- Errors with the instrument itself.
- If the indication checked during preflight is off by more than 75' from the surveyed elevation, the instrument should be referred to a certificated instrument repair station for recalibration.
- Differences between ambient temperature and/or pressure causes an erroneous indication on the altimeter.

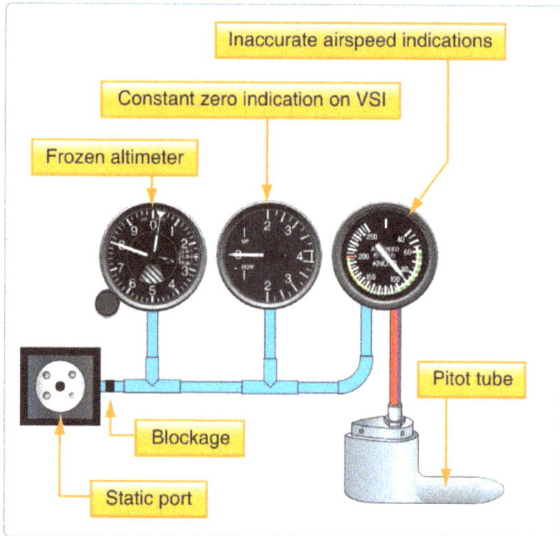


Static Blockage

## Static Port Blockage

- Trapped static pressure causes the altimeter to freeze at the altitude where the blockage occurred.

## Cold Weather Altimeter Errors

A correctly calibrated pressure altimeter indicates true altitude above mean sea level (MSL) when operating within the International Standard Atmosphere (ISA) parameters of pressure and temperature. Nonstandard pressure conditions are corrected by applying the correct local area altimeter setting. Temperature errors from ISA result in true altitude being higher than indicated altitude whenever the temperature is warmer than ISA and true altitude being lower than indicated altitude whenever the temperature is colder than ISA.

True altitude variance under conditions of colder than ISA temperatures poses the risk of inadequate obstacle clearance. Under extremely cold conditions, pilots may need to add an appropriate temperature correction determined from the chart in earlier figures charted IFR altitudes to ensure terrain and obstacle clearance with the following restrictions:

- Altitudes specifically assigned by Air Traffic Control (ATC), such as "maintain 5,000" shall not be corrected. Assigned altitudes may be rejected if the pilot decides that low temperatures pose a risk of inadequate terrain or obstacle clearance.

- If temperature corrections are applied to charted IFR altitudes (such as procedure turn altitudes, final approach fix crossing altitudes, etc), the pilot must advise ATC of the applied correction.

## Altimeter Enhancements (Encoding)

It is not sufficient in the airspace system for only the pilot to have an indication of the aircraft's altitude; the air traffic controller on the ground must also know the altitude of the aircraft. To provide this information, the aircraft is typically equipped with an encoding altimeter.

When the ATC transponder is set to Mode C, the encoding altimeter supplies the transponder with a series of pulses identifying the flight level (in increments of 100') at which the aircraft is flying. This series of pulses is transmitted to the ground radar where they appear on the controller's scope as an alphanumeric display around the return for the aircraft. The transponder allows the ground controller to identify the aircraft and determine the pressure altitude at which it is flying. A computer inside the encoding altimeter measures the pressure referenced from 29.92" Hg and delivers this data to the transponder.

When the pilot adjusts the barometric scale to the local altimeter setting, the data sent to the transponder is not affected. This is to ensure that all Mode C aircraft are transmitting data referenced to a common pressure level. ATC equipment adjusts the displayed altitudes to compensate for local pressure differences allowing display of targets at correct altitudes.

14 CFR part 91 requires the altitude transmitted by the transponder to be within 125' of the altitude indicated on the instrument used to maintain flight altitude.

Table: Lowest flight level correction factor.

| Altimeter Setting | Correction Factor |
|---|---|
| 29.92 or higher | None |
| 29.91 to 29.42 | 500 feet |
| 29.41 to 28.92 | 1000 feet |
| 28.91 to 28.42 | 1500 feet |
| 28.41 to 27.92 | 2000 feet |
| 27.91 to 27.42 | 2500 feet |

## Preflight Action

Should indicate within 75' of field elevation when set with the proper barometric pressure (not regulatory) according to Aeronautical Information Manual 7-2-3. Operating with a known altimeter error, especially in IFR conditions, may not be illegal but certainly falls under careless and reckless operation. If you set your altimeter according to ATIS/AWOS and it is not perfect for field elevation, remember your location on the field.

Filed elevation: the highest elevation of any usable landing surface. Gliders may set the airport elevation so they can know AGL.

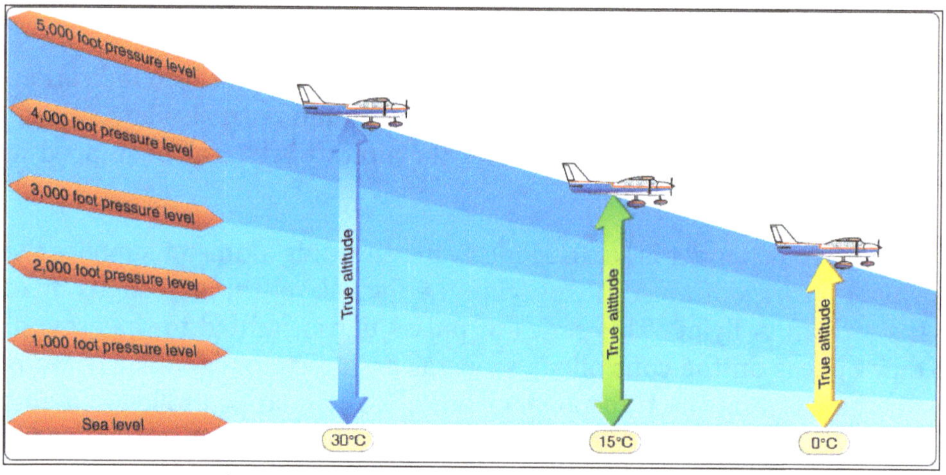


Instrument Flying Handbook, The loss of altitude experienced when flying into an area where the air is colder (more dense) than standard.

Obtaining an altimeter setting:

- Altimeter settings can be obtained from ATC, automated broadcasts, weather reports, as well as various other applications and programs.

## Nonstandard Pressure Effects on an Altimeter

- Maintaining a current altimeter setting is critical because the atmosphere pressure is not constant.
    - That is, in one location the pressure might be higher than the pressure just a short distance away.
- Take an aircraft whose altimeter setting is set to 29.92" of local pressure.
- As the aircraft moves to an area of lower pressure and the pilot fails to readjust the altimeter setting (essentially calibrating it to local pressure), then as the pressure decreases, the true altitude will be lower.
- Adjusting the altimeter settings compensates for this.
- When the altimeter shows an indicated altitude of 5,000', the true altitude at Point A (the height above mean sea level) is only 3,500' at Point B.
- The fact that the altitude indication is not always true lends itself to the memory aid, "When flying from hot to cold or from a high to a low".

## High Barometric Pressure

Cold, dry air masses may produce barometric pressures in excess of 31.00 inches of Mercury, and many altimeters do not have an accurate means of being adjusted for settings of these levels. In this case, the aircraft actual altitude will be higher than the altimeter indicates. When the barometric pressure exceeds 31.00 inches, air traffic controllers will issue the actual altimeter setting, and:

- En Route/Arrivals: Advise pilots to remain set on 31.00 inches until reaching the final approach segment.
- Departures: Advise pilots to set 31.00 inches prior to reaching any mandatory/ crossing altitude or 1,500 feet, whichever is lower.
- The altimeter error caused by the high pressure will be in the opposite direction to the error caused by the cold temperature.

## Low Barometric Pressure

When abnormally low barometric pressure conditions occur (below 28.00), flight operations by aircraft unable to set the actual altimeter setting are not recommended. This is different from high barometric pressure because the true altitude of the aircraft is lower than the indicated altitude if the pilot is unable to set the actual altimeter setting.

## Altimeter Procedures

The cruising altitude or flight level of aircraft must be maintained by reference to an altimeter. Altimeter setting procedures vary based on altitude and atmospheric pressures:

## Operations Below 18,000' MSL

- When barometric pressure is 31.00 in-Hg or less set your altimeter to:
    - To the current reported altimeter setting of a station along the route and within 100 NM of the aircraft, or if there is no station within this area, the current reported altimeter setting of an appropriate available station.
        - When an aircraft is en route on an instrument flight plan, air traffic controllers will furnish this information to the pilot at least once while the aircraft is in the controllers area of jurisdiction.
        - In the case of an aircraft not equipped with a radio, set to the elevation of the departure airport or use an appropriate altimeter setting available prior to departure.
- When barometric pressure exceeds 31.00 in-Hg.
    - The following procedures will be placed in effect by NOTAM defining the geographic area affected:

- For all aircraft: Set 31.00 inches for en route operations below 18,000 feet MSL.
- Maintain this setting until beyond the affected area or until reaching final approach segment.
- At the beginning of the final approach segment, the current altimeter setting will be set, if possible.
- If not possible, 31.00 inches will remain set throughout the approach.
- Aircraft on departure or missed approach will set 31.00 inches prior to reaching any mandatory/crossing altitude or 1,500' AGL, whichever is lower.
    - Air traffic control will issue actual altimeter settings and advise pilots to set 31.00 inches in their altimeters for en route operations below 18,000 feet MSL in affected areas.
- During preflight, barometric altimeters must be checked for normal operation to the extent possible.
- For aircraft with the capability of setting the current altimeter setting and operating into airports with the capability of measuring the current altimeter setting, no additional restrictions apply.
- For aircraft operating VFR, there are no additional restrictions, however, extra diligence in flight planning and in operating in these conditions is essential.
- Airports unable to accurately measure barometric pressures above 31.00 inches of Hg. will report the barometric pressure as "missing" or "in excess of 31.00 inches of Hg":
    - Flight operations to and from those airports are restricted to VFR weather conditions.
- For aircraft operating IFR and unable to set the current altimeter setting, the following restrictions apply:
    - To determine the suitability of departure alternate airports, destination airports, and destination alternate airports, increase ceiling requirements by 100 feet and visibility requirements by 1/4 statute mile for each 1/10 of an inch of Hg., or any portion thereof, over 31.00 inches. These adjusted values are then applied in accordance with the requirements of the applicable operating regulations and operations specifications.
    - Example: Destination altimeter is 31.28 inches, ILS DH 250 feet

(200-1/2). When flight planning, add 300-3/4 to the weather requirements which would become 500-11/4.

- On approach, 31.00 inches will remain set. Decision height (DH) or minimum descent altitude must be deemed to have been reached when the published altitude is displayed on the altimeter.
- Although visibility is normally the limiting factor on an approach, pilots should be aware that when reaching DH the aircraft will be higher than indicated. Using the example above the aircraft would be approximately 300 feet higher.
- These restrictions do not apply to authorized Category II and III ILS operations nor do they apply to certificate holders using approved QFE altimetry systems.

Table: Lowest usable flight level.

| Altimeter | Lowest FL |
|---|---|
| 29.92 or higher | FL 180 |
| 29.91-29.42 | FL 185 |
| 29.41-28.42 | FL 190 |
| 28.91-28.42 | FL 195 |
| 28.41-27.92 | FL 200 |
| 27.91-27.42 | FL 205 |
| 27.41-26.92 | FL 210 |

Table: Altimeter adjustment factor.

| Altimeter Setting | Correction Factor |
|---|---|
| 29.92 or higher | None |
| 29.91 to 29.42 | 500 feet |
| 29.41 to 28.92 | 1000 feet |
| 28.91 to 28.42 | 1500 feet |
| 28.41 to 27.92 | 2000 feet |
| 27.91 to 27.42 | 2500 feet |

## Low Barometric Pressure (Below 28.00 in-Hg)

Flight operations by aircraft unable to set the actual altimeter setting are not recommended. The true altitude of the aircraft is lower than the indicated altitude if the pilot is unable to set the actual altimeter setting.

## Operations at or above 18,000' MSL

- Pilots set the altimeter to the standard 29.92 when entering the "flight levels".

- Since VFR pilots can fly at 17,500' MSL, atmospheric pressures below 29.92 inHg (think "when flying high to low") will bring those pilots flying at 18,000' MSL lower, decreasing the required 500' separation from possible VFR traffic.
- As a result, ATC establishes a lowest usable flight level depending upon the atmospheric pressure in an area of operation.
- The lowest usable flight level is determined by the atmospheric pressure in the area of operation as shown in:
    - Example: The minimum safe altitude of a route is 19,000' MSL and the altimeter setting is reported between 29.92 and 29.42 inches of mercury, the lowest usable flight level will be 195, which is the flight level equivalent of 19,500 feet MSL (minimum altitude plus 500 feet).

## Inertial Reference Unit (IRU), Inertial Navigation System (INS) and Attitude Heading Reference System (AHRS)

IRUs are self-contained systems comprised of gyros and accelerometers that provide aircraft attitude (pitch, roll, and heading), position, and velocity information in response to signals resulting from inertial effects on system components.

- Once aligned with a known position, IRUs continuously calculate position and velocity. IRU position accuracy decays with time.
- This degradation is known as "drift".

INSs combine the components of an IRU with an internal navigation computer.

- By programming a series of waypoints, these systems will navigate along a predetermined track.

AHRSs are electronic devices that provide attitude information to aircraft systems such as weather radar and autopilot, but do not directly compute position information. Aircraft equipped with slaved compass systems may be susceptible to heading errors caused by exposure to magnetic field disturbances (flux fields) found in materials that are commonly located on the surface or buried under taxiways and ramps:

- These materials generate a magnetic flux field that can be sensed by the aircraft's compass system flux detector or "gate", which can cause the aircraft's system to align with the material's magnetic field rather than the earth's natural magnetic field.
- The system's erroneous heading may not self-correct.
- Prior to take off pilots should be aware that a heading misalignment may have occurred during taxi.
- Pilots are encouraged to follow the manufacturer's or other appropriate

procedures to correct possible heading misalignment before take off is commenced.

## Airspeed Indicator

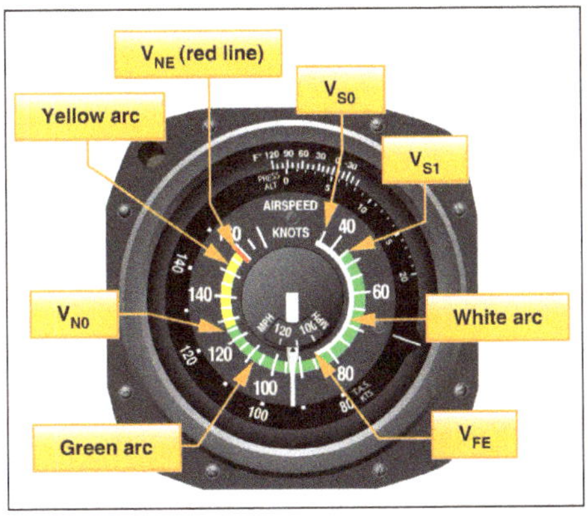

Single-engine ASI (left) with V speeds labelled, and Multi-engine ASI (center) with blue and red radial lines, as well as (right) showing knots only.

The airspeed indicator (ASI) or airspeed gauge is a flight instrument indicating the airspeed of an aircraft in kilometers per hour (km/h), knots (kn), miles per hour (MPH) and/or meters per second (m/s). The recommendation by ICAO is to use km/h, however knots is currently the most used unit. The ASI measures the pressure differential between static pressure from the static port, and total pressure from the pitot tube. This difference in pressure is registered with the ASI pointer on the face of the instrument.

### Usage

The ASI has standard color-coded markings for the safe operation within the limitations of the aircraft. At a glance, the pilot can determine a recommended speed (V speeds) or if he needs to make any speed adjustments. Single and multi-engine aircraft have common markings. For instance, the green arc indicates the normal operating range of the aircraft, from $V_{S1}$ to $V_{NO}$. The white arc indicates the flap operating range, $V_{SO}$ to $V_{FE}$, used for approaches and landings. The yellow arc cautions that flight should be conducted in this range only in smooth air, while the red line ($V_{NE}$) at the top of the yellow arc indicates damage or structural failure may result at higher speeds.

The ASI in multi-engine aircraft includes two additional radial markings, one red and one blue, associated with potential engine failure. The radial red line near the bottom of green arc indicates $V_{mc}$, the minimum indicated airspeed at which the aircraft can be controlled with the critical engine inoperative. The radial blue line indicates $V_{YSE}$, the speed for best rate of climb with the critical engine inoperative.

## Operation

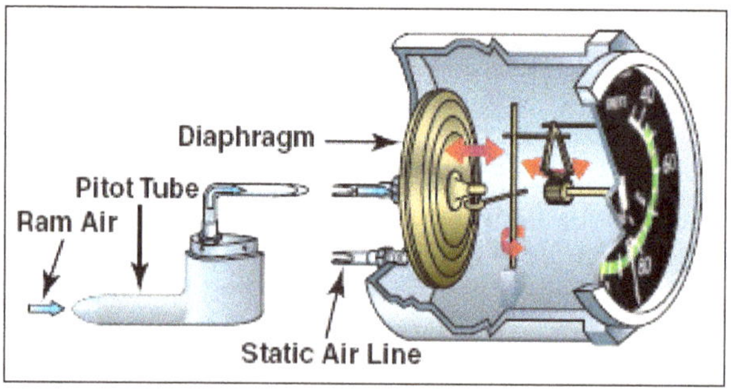


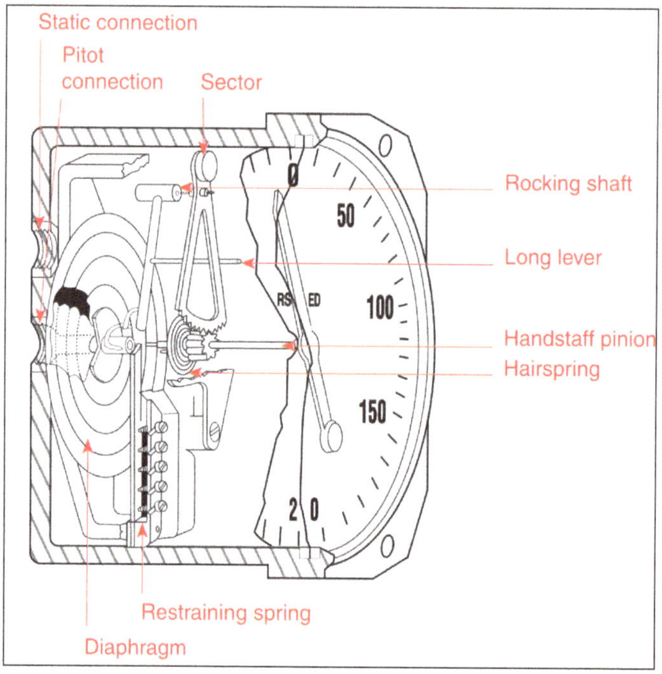


ASI components.

The ASI is the only flight instrument that uses both the static system and the pitot system. Static pressure enters the ASI case, while total pressure flexes the diaphragm, which is connected to the ASI pointer via mechanical linkage. The pressures are equal when the aircraft is stationary on the ground, and hence shows a reading of zero. When the aircraft is moving forward, air entering the pitot tube is at a greater pressure than the static line, which flexes the diaphragm, moving the pointer. The ASI is checked before takeoff for a zero reading, and during takeoff that it is increasing appropriately.

The pitot tube may become blocked, because of insects, dirt or failure to remove the pitot cover. A blockage will prevent ram air from entering the system. If the pitot opening is blocked, but the drain hole is open, the system pressure will drop to ambient pressure, and the ASI pointer will drop to a zero reading. If both the opening and drain holes are blocked, the ASI will not indicate any change in airspeed. However, the ASI pointer will show altitude changes, as the associated static pressure changes. If both the pitot tube and the static system are blocked, the ASI pointer will read zero. If the static ports are blocked but the pitot tube remains open, the ASI will operate, but inaccurately.

## Types of Airspeeds

Left – A mechanical true airspeed indicator for an airplane. The pilot sets the pressure altitude and air temperature in the top window using the knob; the needle indicates true airspeed in the lower-left window. Here the speed is displayed both in knots (kn) and miles per hour (mph). Right – Airspeed indicator with Machmeter, including moveable pointers (*bugs*) at the bezel.

Indicated airspeed (IAS), is read directly off the ASI. It has no correction for air density variations, installation or instrument errors. Calibrated airspeed (CAS) is corrected for installation and instrument errors. An airspeed calibration chart is available for any remaining errors. True airspeed (TAS) is CAS corrected for altitude and nonstandard temperature. TAS is used for flight planning. TAS increases as altitude increases, as air density decreases. TAS may be determined via a flight computer, such as the E6B. Some ASIs have a TAS ring. Alternatively, a rule of thumb is to add 2 percent to the CAS for every 1,000 ft (300 m) of altitude gained.

## Jet Aircraft

Jet aircraft do not have $V_{NO}$ and $V_{NE}$ like piston-engined aircraft, but instead have a maximum operating speed expressed in knots, $V_{MO}$ and Mach number, $M_{MO}$. Thus, a pilot of a jet airplane needs both an airspeed indicator and a Machmeter, with appropriate red lines. An ASI will include a red-and-white striped pointer, or "barber's pole", that automatically moves to indicate the applicable speed limit at any given time.

## Angle of Attack and Lift Reserve Indicators

LRI (left) and AOA Indicators (right)

An airplane can stall at any speed, so monitoring the ASI alone will not prevent a stall. The critical angle of attack (AOA) determines when an aircraft will stall. For a particular configuration, it is a constant independent of weight, bank angle, temperature, density altitude, and the center of gravity of an aircraft. An AOA indicator provides stall situational awareness as a means for monitoring the onset of the critical AOA. The AOA indicator will show the current AOA and its proximity to the critical AOA.

Similarly, the Lift Reserve Indicator (LRI) provides a measure of the amount of lift being generated. It uses a pressure differential system to provide the pilot with a visual representation of reserve lift available.

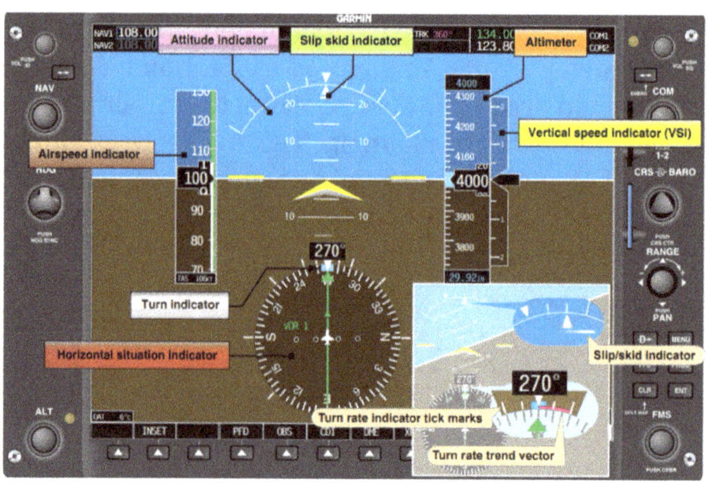

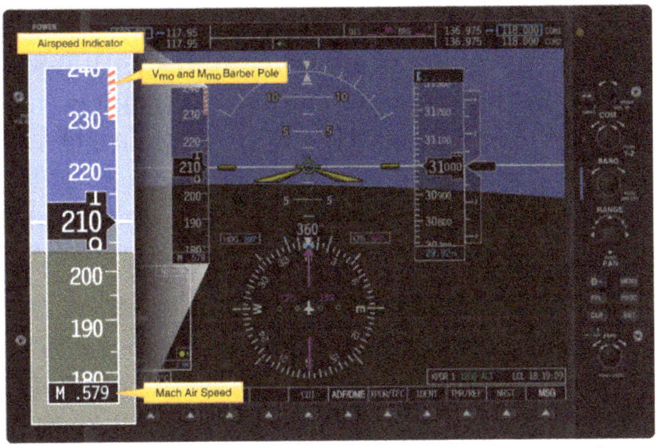


Primary Flight Display (PFD) with the ASI in the form of "Airspeed Tape", including color coded ranges (left), and a combined ASI and Machmeter for a jet aircraft (right).

## Vertical Speed Indicator (Variometer)

A variometer – also known as a rate of climb and descent indicator (RCDI), rate-of-climb indicator, vertical speed indicator (VSI), or vertical velocity indicator (VVI) – is one of the flight instruments in an aircraft used to inform the pilot of the rate of descent or climb. It can be calibrated in feet per minute, knots (1 kn ≈ 100 ft/min) or metres per second (1 m/s ≈ 200 ft/min), depending on country and type of aircraft. It is typically connected to the aircraft's external static pressure source.

In powered flight the pilot makes frequent use of the VSI to ascertain that level flight is being maintained, especially during turning maneuvers. In gliding, the instrument is used almost continuously during normal flight, often with an audible output, to inform the pilot of rising or sinking air. It is usual for gliders to be equipped with more than one type of variometer. The simpler type does not need an external source of power and can therefore be relied upon to function regardless of whether a battery or power source has been fitted. The electronic type with audio needs a power source to be operative during the flight. The instrument is of little interest during launching and landing, with the exception of aerotow, where the pilot will usually want to avoid releasing in sink.

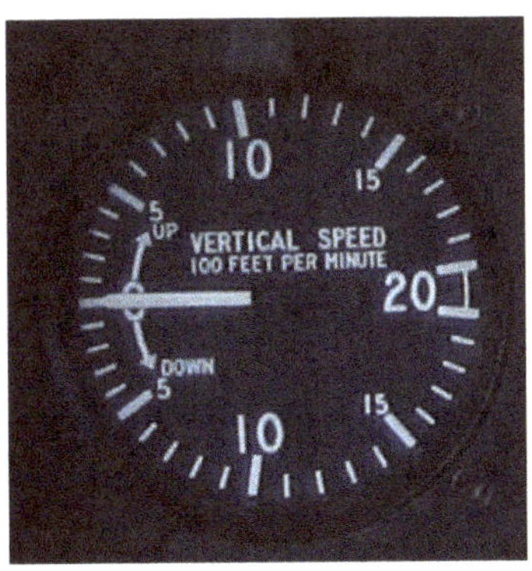


The vertical speed indicator from a Robinson R22. This is the most common type used in aircraft, showing vertical speed in feet per minute (ft/min).

Variometers measure the rate of change of altitude by detecting the change in air pressure (static pressure) as altitude changes. A simple variometer can be constructed by adding a large reservoir (a thermos bottle) to augment the storage capacity of a common aircraft rate-of-climb instrument. In its simplest electronic form, the instrument consists of an air bottle connected to the external atmosphere through a sensitive air flow meter. As the aircraft changes altitude, the atmospheric pressure outside the aircraft changes and air flows into or out of the air bottle to equalise the pressure inside the bottle and outside the aircraft. The rate and direction of flowing air is measured by the cooling of one of two self-heating thermistors and the difference between the thermistor resistances will cause a voltage difference; this is amplified and displayed to the pilot. The faster the aircraft is ascending (or descending), the faster the air flows. Air flowing out of the bottle indicates that the altitude of the aircraft is increasing. Air flowing into the bottle indicates that the aircraft is descending.

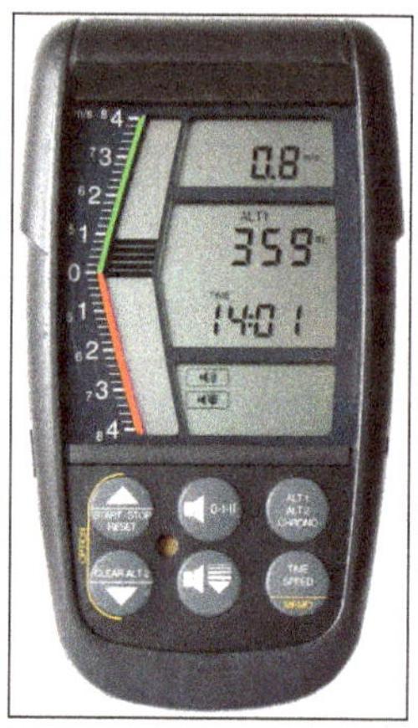

A variometer for paragliders, hang gliders, and ballooneers, showing vertical speed with both a ribbon indicator and a numeric readout, showing vertical speed in meters per second (m/s).

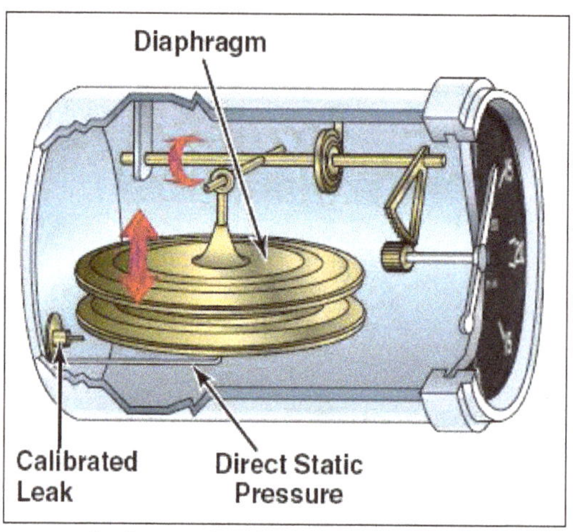


The internals of a classic aircraft vertical speed indicator.

Newer variometer designs directly measure the static pressure of the atmosphere using a pressure sensor and detect changes in altitude directly from the change in air pressure instead of by measuring air flow. These designs tend to be smaller as they do not need the air bottle. They are more reliable as there is no bottle to be affected by changes in temperature and less chances for leaks to occur in the connecting tubes.

The designs described above, which measure the rate of change of altitude by automatically detecting the change in static pressure as the aircraft changes altitude are referred to as "uncompensated" variometers. The term "vertical speed indicator" or "VSI" is most often used for the instrument when it is installed in a powered aircraft. The term "variometer" is most often used when the instrument is installed in a glider or sailplane.

An "Inertial-lead" or "Instantaneous" VSI (IVSI) uses accelerometers to provide a quicker response to changes in vertical speed.

Panel mounted variometer for gliders, showing vertical speed in knots (kn).

## Purpose

Human beings, unlike birds and other flying animals, are not able directly to sense climb and sink rates. Before the invention of the variometer, sailplane pilots found it very hard to soar. Although they could readily detect abrupt *changes* in vertical speed ("in the seat of the pants"), their senses did not allow them to distinguish lift from sink, or strong lift from weak lift. The *actual* climb/sink rate could not even be guessed at, unless there was some clear fixed visual reference nearby. Being near a fixed reference means being near to a hillside, or to the ground. Except when hill-soaring (exploiting the lift close to the up-wind side of a hill), these are generally very unprofitable positions for glider pilots to be in. The most useful forms of lift (thermal and wave lift) are found at higher altitudes and it is very hard for a pilot to detect or exploit them without the use of a variometer. After the variometer was invented in 1929 by Alexander Lippisch and Robert Kronfeld, the sport of gliding moved into a new realm.

Variometers also became important in foot-launch hang gliding, where the open-to-air pilot hears the wind but needs the variometer to help him or her to detect regions of rising or sinking air. In early hang gliding, variometers were not needed for the short flights or flights close to ridge lift. But the variometer became key as pilots began making longer flights. The first portable variometer for use in hang gliders was the Colver Variometer by Colver Soaring Instruments which served to extend the sport into cross-country thermal flying.

## Total Energy Compensation

As the sport of gliding developed, however, it was found that these very simple "uncompensated" instruments had their limitations. The information that glider pilots really need to soar is the total change in energy experienced by the glider, including both altitude and speed. An uncompensated variometer will simply indicate vertical speed of the glider, giving rise to the possibility of a "stick thermal," i.e., a change in altitude caused by stick input only. If a pilot pulls back on the stick, the glider will rise, but also slow down as well. But if a glider is rising without the speed changing, this is an indication of real lift, not "stick lift."

Compensated variometers also include information about the speed of the aircraft, so the total energy (potential and kinetic) is used, not just the change in altitude. For example, if a pilot pushes forward on the stick, speeding up as the plane dives, an uncompensated variometer only indicates that altitude is being lost. But the pilot could pull back on the stick, trading the extra speed for altitude again. A compensated variometer uses both speed and altitude to indicate the change in total energy. So the pilot that pushes the stick forward, diving to gain speed, and then pulls back again to regain altitude will notice no change in total energy on a compensated variometer (neglecting energy loss due to drag).

Most modern sailplanes are equipped with *Total Energy compensated* variometers.

## Total Energy Compensation in Theory

The total energy of the aircraft is:

- $E_{tot} = E_{pot} + E_{kin}$

  where $E_{pot}$ is the potential energy, and $E_{kin}$ is the kinetic energy. So the change in total energy is:

- $\Delta E_{tot} = \Delta E_{pot} + \Delta E_{kin}$

The Vertical Speed Indicator in this Van's Aircraft RV-4 light aircraft is within the yellow rectangle.

Since,

- Potential Energy is proportional to Height,

  $$E_{pot} = mgh$$

  Where, m is the glider mass and g the acceleration of gravity.

- Kinetic Energy is proportional to Velocity squared,

  $$E_{kin} = \frac{1}{2}mV^2$$

  then from 2:

- $\Delta E_{tot} = mg\Delta h + \frac{1}{2}m\Delta V^2$
- Typically, this is converted to an effective altitude change by dividing by the acceleration of gravity, and the mass of the aircraft, so:

  $$\frac{E_{tot}}{mg} = \Delta h + \frac{\Delta V^2}{2g}$$

## Total Energy Compensation in Practice

In most sailplanes, total energy compensation is achieved by connecting the variometer to the atmosphere via a "total energy probe", that produces vacuum proportional to the square of the glider's air speed—in effect, a negative pitot. Alternatively, the subtraction may be done electronically by the flight computer based on indicated airspeed (pitot).

Very few powered aircraft have total energy variometers. Pilots of powered aircraft are more interested in the true rate of change of altitude, as they often want to hold a constant altitude or maintain a steady climb or descent.

The total energy probe used to be shaped as a classical venturi (two small funnels connected back-to-back by their narrow ends), or nowadays the Irving Tube—a slot or pair of holes on the back side of a quarter inch vertical tube. The geometry of the total energy probe is such that air flow generates suction (reduced pressure).

To maximise the precision of this compensation effect, the total energy probe needs to be in undisturbed airflow ahead of the aircraft nose or tail fin (the "Braunschweig tube",the long cantilevered tube with a kink in the end that can be seen projecting from the leading edge of the tail fin on most modern sailplanes).

The ideal location for a total energy variometer is a place in the flow field around the aircraft where the pressure coefficient is negative one.

## Netto Variometer

A second type of compensated variometer is the Netto or airmass variometer. In addition to TE compensation, the Netto variometer adjusts for the intrinsic sink rate of the glider at a given speed (the polar curve) adjusted for the wing loading due to water ballast. The Netto variometer will always read zero in still air. This provides the pilot with the accurate measurement of air mass vertical movement critical for final glides (the last glide to the ultimate destination location).

The Relative Netto Variometer indicates the vertical speed the glider would achieve IF it flies at thermalling speed - independent of current air speed and attitude. This reading is calculated as the Netto reading minus the glider's minimum sink.

When the glider circles to thermal, the pilot needs to know the glider's vertical speed instead of that of the air mass. The Relative Netto Variometer (or sometimes the super Netto) includes a g-sensor to detect thermalling.

When thermalling, the sensor will detect acceleration (gravity plus centrifugal) above 1 g and tell the relative netto variometer to stop subtracting the sailplane's wing load-adjusted polar sink rate for the duration. Some earlier nettos used a manual switch instead of the g sensor.

## Electronic Variometers

In modern gliders, most electronic variometers generate a sound whose pitch and rhythm depends on the instrument reading. Typically the audio tone increases in frequency as the variometer shows a higher rate of climb and decreases in frequency towards a deep groan as the variometer shows a faster rate of descent. When the variometer is showing a climb, the tone is often chopped and the rate of chopping may be increased as the climb rate increases, while during a descent the tone is not chopped. The vario is typically silent in still air or in lift which is weaker than the typical sink rate of the glider at minimum sink.

This audio signal allows the pilot to concentrate on the external view instead of having to watch the instruments, thus improving safety and also giving the pilot more opportunity to search for promising looking clouds and other signs of lift. A variometer that produces this type of audible tone is known as an "audio variometer".

Advanced electronic variometers in gliders can present other information to the pilot from GPS receivers. The display can thus show the bearing, distance and height required to reach an objective. In cruise mode (used in straight flight), the vario can also give an audible indication of the correct speed to fly depending on whether the air is rising or sinking. The pilot merely has to input the estimated MacCready setting, which is the expected rate of climb in the next acceptable thermal.

There is an increasing trend for advanced variometers in gliders towards flight computers (with variometer indications) which can also present information such as controlled airspace, lists of turnpoints and even collision warnings. Some will also store positional GPS data during the flight for later analysis.

## Radio Controlled Soaring

Variometers are also used in radio controlled gliders. Each variometer system consists of a radio transmitter in the glider, and a receiver on the ground for use by the pilot. Depending on the design, the receiver may give the pilot the current altitude of the glider, and a display that indicates if the glider is gaining or losing altitude—often via an audio tone. Other forms of telemetry may also be provided by the system, displaying parameters such as airspeed and battery voltage. Variometers used in radio controlled gliders may or may not feature total energy compensation.

Variometers are not essential in radio controlled gliders; a skilled pilot can usually determine if the glider is going up or down via visual cues alone. The use of variometers is prohibited in some soaring contests for radio controlled gliders.

## Machmeter

A Machmeter is an aircraft pitot-static system flight instrument that shows the ratio of the true airspeed to the speed of sound, a dimensionless quantity called Mach number. This is shown on a Machmeter as a decimal fraction. An aircraft flying at the speed of sound is flying at a Mach number of one, expressed as *Mach 1*.

As an aircraft in transonic flight approaches the speed of sound, it first reaches its critical mach number, where air flowing over low-pressure areas of its surface locally reaches the speed of sound, forming shock waves. The indicated airspeed for this condition changes with ambient temperature, which in turn changes with altitude. Therefore, indicated airspeed is not entirely adequate to warn the pilot of the impending problems. Mach number is more useful, and most high-speed aircraft are limited to a maximum operating Mach number, also known as $M_{MO}$.

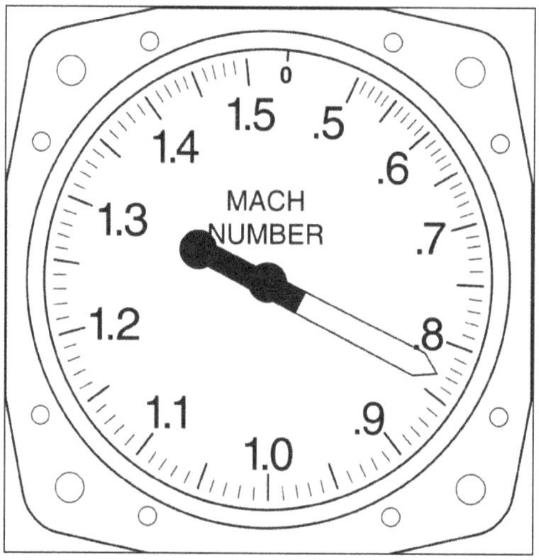


Illustration showing the face of a Machmeter reading a Mach number of 0.83.

For example, if the $M_{MO}$ is Mach 0.83, then at 9,100 m (30,000 ft) where the speed of sound under standard conditions is 1,093 kilometres per hour (590 kn), the true airspeed at $M_{MO}$ is 906 kilometres per hour (489 kn). The speed of sound increases with air temperature, so at Mach 0.83 at 3,000 m (10,000 ft) where the air is much warmer than at 9,100 m (30,000 ft), the true airspeed at $M_{MO}$ would be 982 km/h (530 kn).

## Operation

Modern electronic Machmeters use information from an air data computer system which makes calculations using inputs from a pitot-static system. Some older mechanical Machmeters use an altitude aneroid and an airspeed capsule which together convert pitot-static pressure into Mach number. The Machmeter suffers from instrument and position errors.

## Calibration

In subsonic flow the Mach meter can be calibrated according to:

$$M = \sqrt{5\left[\left(\frac{q_c}{p}+1\right)^{\frac{2}{7}}-1\right]} or, M = \sqrt{5\left[\left(\frac{p_t}{p}\right)^{\frac{2}{7}}-1\right]}$$

Where:

- $M$ is Mach number.
- $q_c$ is impact pressure (dynamic pressure).
- $p$ is static pressure.
- And assuming the ratio of specific heats is 1.4.

When a shock wave forms across the pitot tube the required formula is derived from the Rayleigh Supersonic Pitot equation, and is solved iteratively:

$$M = 0.88128485\sqrt{\frac{p_t}{p}\left(1-\frac{1}{[7M^2]}\right)^{\frac{5}{2}}}$$

where:

- $p_t$ is now total pressure measured behind the normal shock.

Note that the inputs required are total pressure and static pressure. Air temperature input is not required.

# GYROSCOPIC SYSTEMS

Most aircraft have several instruments that are traditionally driven by mechanical Gyros. These instruments assist in flying and navigation of an aircraft. These instruments are the Attitude Indicator (also known as the Vertical Gyro), the Directional Gyro, and the Turn and Bank Indicator. Aircraft also typically have a compass, and in some cases a Flux Valve (also known as a Magnetometer) to which the Directional Gyro is connected or slaved to cancel long term drift. If the aircraft does not have an electronic Flux Valve, then the Directional Gyro or DG has to be manually reset to the compass reading during straight and level flight (when the compass is accurate) on a periodic basis. In most light aircraft the Turn Coordinator (TC) is electrically driven. Usually the Heading Indicator (HI) and Attitude Indicator (AI) are vacuum driven.

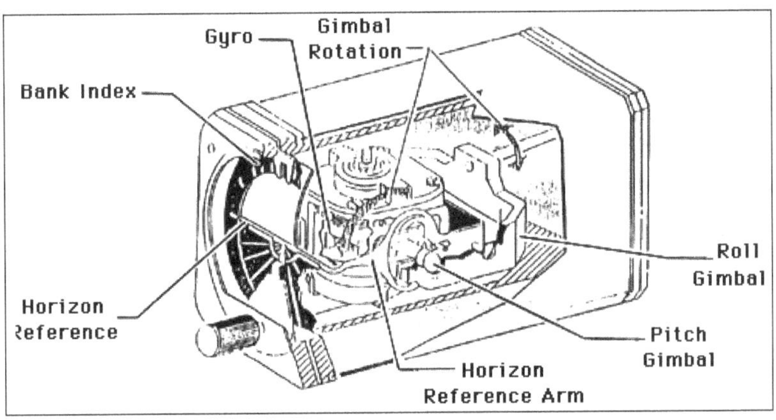


The three Gyro instruments, Attitude Indicator, Directional Gyro and Turn and Bank Indicator are 'gyro' driven. What does gyro-driven mean? A gyro is a spinning wheel (mass) that obeys the Laws of Physics. The spinning wheel is spun up either electrically (electric gyros) or via air flow (vacuum gyros) to high rotational speeds and a high angular momentum. The spinning wheel is mechanically isolated from the casing of the instrument thru a series of gimbals. Due to the conservation of angular momentum, the spinning wheel will try to maintain its orientation, via the gimbals, as the outer casing moves. The outer casing is of course connected to the airframe. The gimbals, move by the amount the aircraft has rolled, pitched, or changed heading, and in some cases directly connect to the display. The display provides an indication of the aircraft attitude. In the case of a remote gyro and also with many electric gyros, the gimbals provide an analogue electrical output proportional to aircraft orientation change.

## Gyroscopic Principles

Any spinning object possesses gyroscopic characteristics. The central mechanism of the gyroscope is a wheel similar to a bicycle wheel. It's outer rim has a heavy mass. It rotates at high speed on very low friction bearings. When it is rotating normally, it resists changes in direction.

The gyroscope exhibits two predominant characteristics:

- Rigidity in Space.
- Precession.

## Rigidity in Space

The gyroscope resists turning. When it is "gimballed" ( free to move in a given direction) such that it is free to move either in 1, 2 or 3 dimensions, any surface such as an instrument dial attached to the gyro assembly will remain rigid in space even though the case of the gyro turns. The Attitude Indicator (AI) and the Heading Indicator (HI ) use this property of rigidity in space for their operation. The HI responds only to change of heading. The AI responds to both changes in Pitch and in Roll.

## Precession

Precession is the deflection of a spinning wheel 90° to the plane of rotation when a deflective force is applied at the rim. If a force is applied the top of the rim (the plane of rotation), the precession (turn) will be 90° in the horizontal plane to the left. The Turn Coordinator (TC) uses this precession property. For example, then taxiing on the ground, the Turn Coordinator will move, with the small airplane in the instrument showing a bank, even though the aircraft is level. The banking of the small aircraft presentation indicates only that the aircraft is turning.

## The Vacuum System

The Attitude Indicator (AI) and the Heading Indicator (HI) in light aircraft are usually driven by a vacuum system. The principal components are shown below. Not shown are auxiliary devices such as valves, filters etc. A pump provides the vacuum to the AI and HI through a system of vacuum lines. A Vacuum Gauge is attached to the lines which gives the pilot an indication that adequate vacuum is being generated.

## Gyro Power Sources

Air or electricity supply the power to operate gyro instruments in light aircraft. If the directional indicator and attitude indicator are air-driven (as they generally are), the turn-and-slip indicator is electrically powered. The advantage of this arrangement is that if the vacuum system (which supplies air) fails, the instrument pilot still has the compass and the turn indicator for attitude and direction reference, in addition to the pitot-static instruments.

1. Vacuum power system: Air-driven gyros normally are powered by a vacuum pump attached to and driven by the engine. Suction lines connect the pump to the instruments, drawing cabin air through the filtered openings in the instrument case. As the air enters the case, it is accelerated and directed against small "buckets" cast into the gyro wheel. A regulator is attached between the pump and the gyro instrument case to control suction pressure. There is normally a vacuum gauge, suction gauge or warning light. Because a constant gyro speed is essential for reliable instrument readings, the correct suction pressure is maintained with a vacuum pressure regulator.

Typical suction gauge.

The air is drawn through a filter, to the instruments and then to the pump where it is vented to atmosphere. The pilot should consult the aircraft operating manual for specific information with regard to vacuum system normal operating values. Low gyro rotation speeds cause slow instrument response or lagging indications, while fast gyro speeds cause the instruments to overreact in addition to wearing the gyro bearings faster and decreasing gyro life.

2. Electrical power system: An electric gyro, normally used to drive the turn coordinator or turn-and-slip indicator, operates like a small electric motor with the spinning gyro acting as the motor armature. Gyro speed in these instruments is approximately 8,000 rpm.

Aircraft that normally operate at high altitudes do not use a vacuum system to power flight instruments because pump efficiency is limited in the thin, cold air. Instead, alternating current (a.c.) drives the gyros in the heading and attitude indicators. The a.c. power is provided by inverters that convert direct current to alternating current. In some cases, the a.c. power is supplied directly from the engine-driven alternator or generator.

## Attitude Indicator

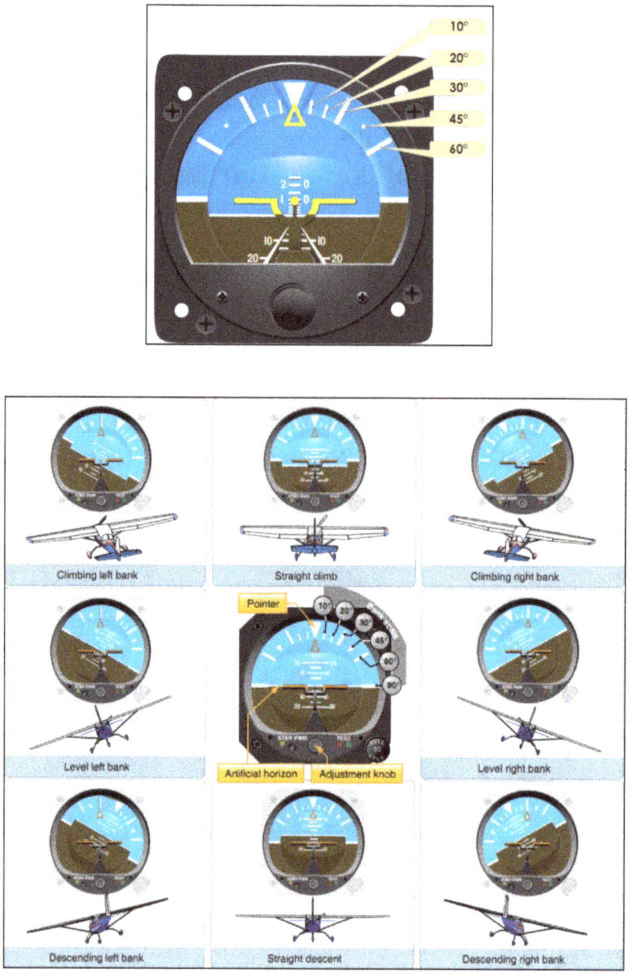


AI with pitch and roll reference lines (top) and the AI relationship to aircraft orientation (down).

The attitude indicator (AI), formerly known as the gyro horizon or artificial horizon, is a flight instrument that informs the pilot of the aircraft orientation relative to Earth's horizon, and gives an immediate indication of the smallest orientation change. The miniature aircraft and horizon bar mimic the relationship of the aircraft relative to the actual horizon. It is a primary instrument for flight in instrument meteorological conditions.

Attitude is always presented to users in the unit degrees (°). However, inner workings such as sensors, data and calculations may use a mix of degrees and radians, as scientists and engineers may prefer to work with radians.

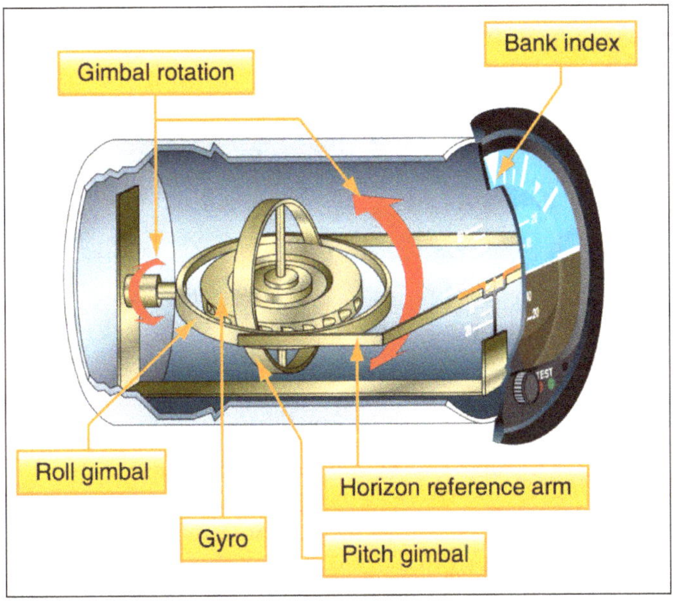


AI interior.

The essential components of the AI include a symbolic miniature aircraft mounted so that it appears to be flying relative to the horizon. An adjustment knob, to account for the pilot's line of vision, moves the aircraft up and down to align it against the horizon bar. The top half of the instrument is blue to represent the sky, while the bottom half is brown to represent the ground. The bank index at the top shows the aircraft angle of bank. Reference lines in the middle indicate the degree of pitch, up or down, relative to the horizon.

Most Russian-built aircraft have a somewhat different design. The background display is colored as in a Western instrument, but moves up and down only to indicate pitch. A symbol representing the aircraft (which is fixed in a Western instrument) rolls left or right to indicate bank angle. A proposed hybrid version of the Western and Russian systems that would be more intuitive, never caught on.

## Operation

The heart of the AI is a gyroscope (gyro) that spins at high speed, from either an electric motor, or through the action of a stream of air pushing on rotor vanes placed along its periphery. The stream of air is provided by a vacuum system, driven by a vacuum pump, or a venturi. Air passing through the narrowest portion of a venturi has lower air pressure through Bernoulli's Principle. The gyro is mounted in a double gimbal, which allows the aircraft to pitch and roll as the gyro stays vertically upright. A self-erecting mechanism, actuated by gravity, counteracts any precession due to bearing friction. It may take a few minutes for the erecting mechanism to bring the gyros to a vertical upright position after the aircraft engine is first powered up.

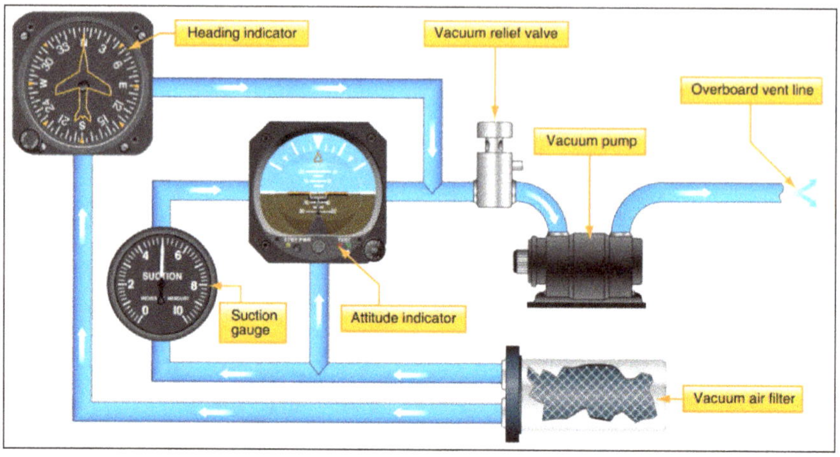


Vacuum system using a vacuum pump.

Attitude indicators have mechanisms that keep the instrument level with respect to the direction of gravity. The instrument may develop small errors, in pitch or bank during extended periods of acceleration, deceleration, turns, or due to the earth curving underneath the plane on long trips. To start with, they often have slightly more weight in the bottom, so that when the aircraft is resting on the ground they will hang level and therefore they will be level when started. But once they are started, that pendulous weight in the bottom will not pull them level if they are out of level, but instead its pull will cause the gyro to precess. In order to let the gyro very slowly orient itself to the direction of gravity while in operation, the typical vacuum powered gyro has small pendulums on the rotor casing that partially cover air holes. When the gyro is out of level with respect to the direction of gravity, the pendulums will swing in the direction of gravity and either uncover or cover the holes, such that air is allowed or prevented from jetting out of the holes, and thereby applying a small force to orient the gyro towards the direction of gravity. Electric powered gyros may have different mechanisms to achieve a similar effect.

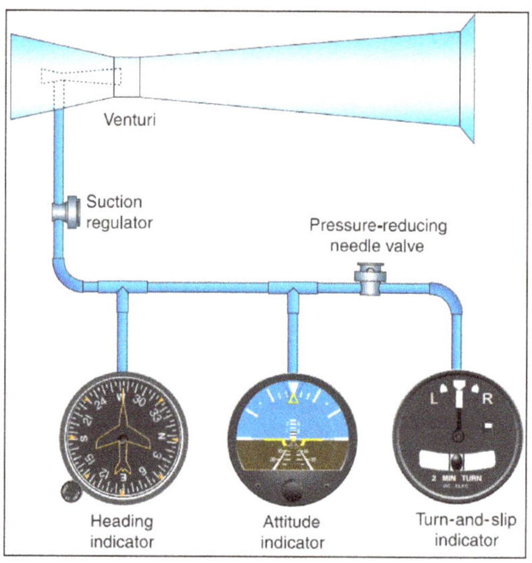


Vacuum system using a venture.

Older AIs were limited in the amount of pitch or roll that they would tolerate. Exceeding these limits would cause the gyro to tumble as the gyro housing contacted the gimbals, causing a precession force. Preventing this required a caging mechanism to lock the gyro if the pitch exceed 60° and the roll exceeded 100°. Modern AIs don't have this limitation and don't require a caging mechanism.

## Flight Director Attitude Indicator

Attitude indicators are also used on manned spacecraft and are called Flight Director Attitude Indicators (FDAI), where they indicate the craft's yaw angle (nose left or right), pitch (nose up or down), roll, and orbit relative to a fixed-space inertial reference frame from an Inertial Measurement Unit (IMU). The FDAI can be configured to use known positions relative to Earth or the stars, so that the engineers, scientists and astronauts can communicate the relative position, attitude, and orbit of the craft.

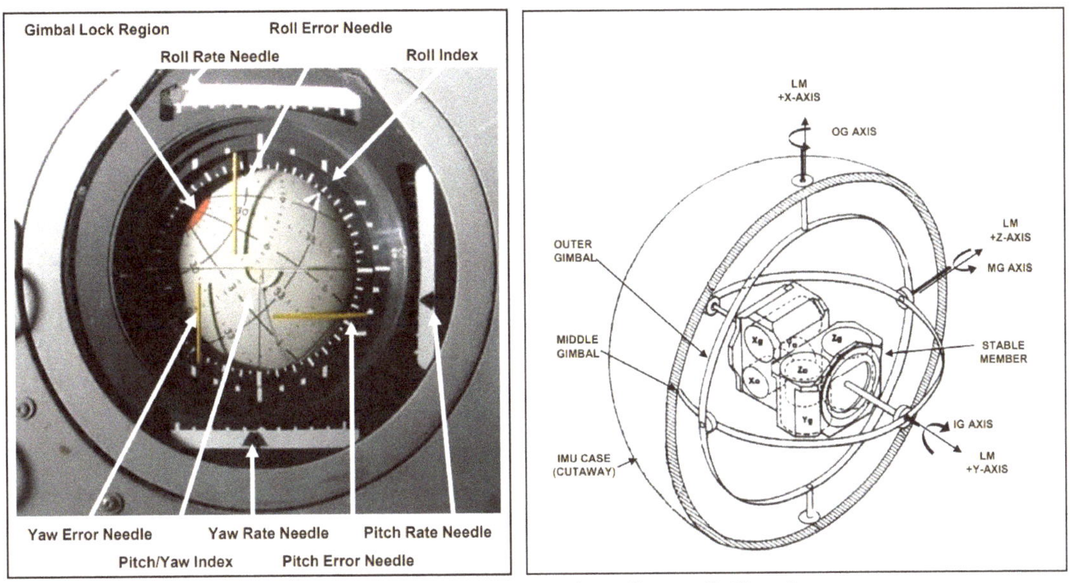


Apollo Flight Director Attitude Indicator (left) and
Inertial Measurement Unit (IMU) (right).

## Attitude and Heading Reference Systems

Attitude and Heading Reference Systems (AHRS) are able to provide three-axis information based on ring laser gyroscopes, that can be shared with multiple devices in the aircraft, such as "glass cockpit" primary flight displays (PFDs). Rather than using a spinning gyroscope, modern AHRS use solid-state electronics, low-cost inertial sensors, rate gyros, and magnetometers.

With most AHRS systems, if an aircraft's AIs have failed there will be a standby AI located in the center of the instrument panel, where other standby basic instruments such as the airspeed indicator and altimeter are also available. These mostly mechanical standby instruments may be available even if the electronic flight instruments fail, though the standby attitude indicator may be electrically driven and will, after a short time, fail if its electrical power fails.

## Attitude Direction Indicator

ADI (left) with yellow V steering bars, and an AI integrated with ILS glide slope and localizer indicators (right).

The Attitude Direction Indicator (ADI), or Flight Director Indicator (FDI), is an AI integrated with a Flight Director System (FDS). The ADI incorporates a computer that receives information from the navigation system, such as the AHRS, and processes this information to provide the pilot with a 3-D flight trajectory cue to maintain a desired path. The cue takes the form of V steering bars. The aircraft is represented by a delta symbol and the pilot flies the aircraft so that the delta symbol is placed within the V steering bars.

Soviet artificial horizon AGP-2, tilted left, and showing the aircraft nose down and banked to the left. The white "horizon" line always aligns with the wings, not with the horizon seen out of the cockpit.

## Heading Indicator

A heading indicator in a small aircraft.

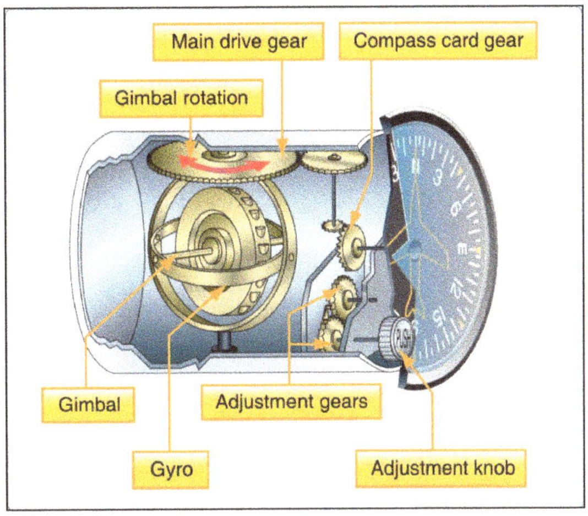


HI interior.

The heading indicator (also called an HI) is a flight instrument used in an aircraft to inform the pilot of the aircraft's heading. It is sometimes referred to by its older names, the directional gyro or DG, and also (UK usage) direction indicator or DI.

The primary means of establishing the heading in most small aircraft is the magnetic compass, which, however, suffers from several types of errors, including that created by the "dip" or downward slope of the Earth's magnetic field. Dip error causes the magnetic compass to read incorrectly whenever the aircraft is in a bank, or during acceleration or deceleration, making it difficult to use in any flight condition other than unaccelerated, perfectly straight and level. To remedy this, the pilot will typically manoeuvre the airplane with reference to the heading indicator, as the gyroscopic heading indicator is unaffected by dip and acceleration errors. The pilot will periodically reset the heading indicator to the heading shown on the magnetic compass.

### Operation

The heading indicator works using a gyroscope, tied by an erection mechanism to the aircraft yawing plane, i. e. the plane defined by the longitudinal and the transverse axis of the aircraft. As such, any configuration of the aircraft yawing plane that does not match the local Earth horizontal results in an indication error. The heading indicator is arranged such that the gyro axis is used to drive the display, which consists of a circular compass card calibrated in degrees. The gyroscope is spun either electrically, or using filtered air flow from a suction pump (sometimes a pressure pump in high altitude aircraft) driven from the aircraft's engine. Because the Earth rotates (ω, 15° per hour, apparent drift), and because of small accumulated errors caused by imperfect balancing of the gyro, the heading indicator will drift over time (real drift), and must be reset using a magnetic compass periodically. The apparent drift is predicted by ω sin Latitude and will thus be greatest over the poles. To counter for the effect of Earth rate drift a latitude nut can be set (on the ground only) which induces a (hopefully equal and opposite) real wander in the gyroscope. Otherwise it would be necessary to manually realign the direction indicator once each ten to fifteen minutes during routine in-flight checks. Failure to do this is a common source of navigation errors among new pilots. Another sort of apparent drift exists in the form of transport wander, caused by the aircraft movement and the convergence of the meridian lines towards the poles. It equals the course change along a great circle (orthodrome) flight path.

## Variations

Some more expensive heading indicators are "slaved" to a magnetic sensor, called a "flux gate". The flux gate continuously senses the Earth's magnetic field, and a servo mechanism constantly corrects the heading indicator. These "slaved gyros" reduce pilot workload by eliminating the need for manual realignment every ten to fifteen minutes.

The prediction of drift in degrees per hour, is as follows:

| SOURCE | Drift Rate (Degrees per Hour) | Sign (Northern Hemisphere) | Sign (Southern Hemisphere) |
|---|---|---|---|
| Earth Rate | 15 * sin Operating Latitude | - (causing an under-read) | + (causing an over-read) |
| Latitude Nut | 15 * sin Latitude of Setting | + | - |
| Transport Wander EAST | (East Ground Speed Component (or sine track angle x ground speed or change in longitude/flight time in hours) * tan Operating Latitude)/60 | - | + |
| Transport Wander WEST | (West Ground Speed Component (or sine track angle x ground speed or change in longitude/flight time in hours) * tan Operating Latitude)/60 | + | - |
| Real/Random Wander | As given in the Aircraft Operating Manual | As given | As given |

Although it is possible to predict the drift, there will be minor variations from this basic model, accounted for by gimbal error (operating the aircraft away from the local horizontal), among others. A common source of error here is the improper setting of the latitude nut (to the opposite hemisphere for example). The table however allows one to gauge whether an indicator is behaving as expected, and as such, is compared with the realignment corrections made with reference to the magnetic compass. Transport wander is an undesirable consequence of apparent drift.

## Turn and Slip Indicator

In aviation, the turn and slip indicator (T/S, a.k.a. turn and bank indicator) and the turn coordinator (TC) variant are essentially two aircraft flight instruments in one device. One indicates the rate of turn, or the rate of change in the aircraft's heading; the other part indicates whether the aircraft is in coordinated flight, showing the slip or skid of the turn. The slip indicator is actually an inclinometer that at rest displays the angle of the aircraft's transverse axis with respect to horizontal, and in motion displays this angle as modified by the acceleration of the aircraft. The most commonly used units are degrees per second (deg/s) or minutes per turn (min/tr).

Illustration of the face of a turn-and-slip indicator.

## Name

The turn and slip indicator can be referred to as the turn and bank indicator, although the instrument does not respond directly to bank angle. Neither does the turn coordinator, but it does respond to roll rate, which enables it to respond more quickly to the start of a turn.

## Operation

## Turn Indicator

The turn indicator is a gyroscopic instrument that works on the principle of precession. The gyro is mounted in a gimbal. The gyro's rotational axis is in-line with the transverse (pitch) axis of the aircraft, while the gimbal has limited freedom around the longitudinal (roll) axis of the aircraft.

As the aircraft yaws, a torque force is applied to the gyro around the vertical axis, due to aircraft yaw, which causes gyro precession around the roll axis. The gyro spins on an axis that is 90 degrees relative to the direction of the applied yaw torque force. The gyro and gimbal rotate (around the roll axis) with limited freedom against a calibrated spring.

The torque force against the spring reaches an equilibrium and the angle that the gimbal and gyro become positioned is directly connected to the display needle, thereby indicating the rate of turn. In the turn coordinator, the gyro is canted 30 degrees from the horizontal so it responds to roll as well as yaw.

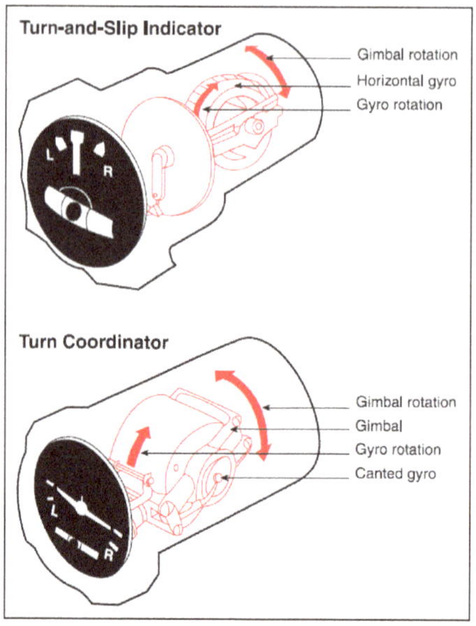


Graphic of a turn and slip indicator and a turn coordinator.

The display contains hash marks for the pilot's reference during a turn. When the needle is lined up with a hash mark, the aircraft is performing a "standard rate turn" which is defined as three degrees per second, known in some countries as "rate one". This translates to two minutes per 360 degrees of turn (a complete circle). Indicators are marked as to their sensitivity, with "2 min turn" for those whose hash marks correspond to a standard rate or two-minute turn, and "4 min turn" for those, used in faster aircraft, that show a half standard rate or four-minute turn. The supersonic Concorde jet aircraft and many military jets are examples of aircraft that use 4 min. turn indicators. The hash marks are sometimes called "dog houses", because of their distinct shape on various makes of turn indicators. Under instrument flight rules, using these figures allows a pilot to perform timed turns in order to conform with the required air traffic patterns. For a change of heading of 90 degrees, a turn lasting 30 seconds would be required to perform a standard rate or "rate one" turn.

## Inclinometer

Coordinated flight indication is obtained by using an inclinometer, which is recognized as the "ball in a tube". An inclinometer contains a ball sealed inside a curved glass tube, which also contains a liquid to act as a damping medium. The original form of the indicator is in effect a spirit level with the tube curved in the opposite direction and a bubble replacing the ball. In some early aircraft the indicator was merely a pendulum with a dashpot for damping. The ball gives an indication of whether the aircraft is slipping, skidding or in coordinated flight. The ball's movement is caused by the force of gravity and the aircraft's centripetal acceleration. When the ball is centered in the middle of the tube, the aircraft is said to be in coordinated flight. If the ball is on the inside (wing down side) of a turn, the aircraft is slipping. And finally, when the ball is on the outside (wing up side) of the turn, the aircraft is skidding.

A simple alternative to the balance indicator used on gliders is a yaw string, which allows the pilot to simply view the string's movements as rudimentary indication of aircraft balance.

## Turn Coordinator

Image showing the face of a turn coordinator during a standard rate coordinated right turn.

The turn coordinator (TC) is a further development of the turn and slip indicator (T/S) with the major difference being the display and the axis upon which the gimbal is mounted. The display is that of a miniature airplane as seen from behind. This looks similar to that of an attitude indicator. "No pitch information" is usually written on the instrument to avoid confusion regarding the aircraft's pitch, which can be obtained from the artificial horizon instrument.

In contrast to the T/S, the TC's gimbal is pitched up 30 degrees from the transverse axis. This causes the instrument to respond to roll as well as yaw. This allows the instrument to display a change more quickly as it will react to the change in roll before the aircraft has even begun to yaw. Although this instrument reacts to changes in the aircraft's roll, it does not display the roll attitude.

The turn coordinator may be used as a performance instrument when the attitude indicator has failed. This is called "partial panel" operations. It can be unnecessarily difficult or even impossible if the pilot does not understand that the instrument is showing roll rates as well as turn rates. The usefulness is also impaired if the internal dashpot is worn out. In the latter case, the instrument is underdamped and in turbulence will indicate large full-scale deflections to the left and right, all of which are actually roll rate responses.

## Practical Implications

Slipping and skidding within a turn is sometimes referred to as a sloppy turn, due to the perceptive discomfort it can cause to the pilot and passengers. When the aircraft is in a balanced turn (ball is centered), passengers experience gravity directly in line with their seat (force perpendicular to seat). With a well balanced turn, passengers may not even realize the aircraft is turning unless they are viewing objects outside the aircraft.

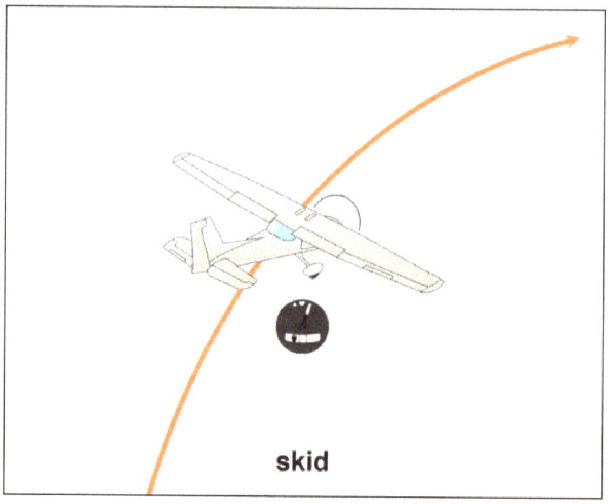


Skidding: The corrective action is to press the "left" pedal adequately.

While aircraft slipping and skidding are often undesired in a usual turn that maintains altitude, slipping of the aircraft can be used for practical purposes. Intentionally putting an aircraft into a slip is used as a forward slip and a sideslip. These slips are performed by applying opposite inputs of the aileron and rudder controls. A forward slip allows a pilot to quickly drop altitude without gaining unnecessary speed, while a sideslip is one method utilized to perform a crosswind landing.

## References


- Pitot-Static-System: skybrary.aero, Retrieved 19 August, 2020
- Altimeter, technology: britannica.com, Retrieved 15 May, 2020
- A. C. Kermode (2006). Mechanics of Flight. Pitman Boots LTD. p. 243. ISBN 978-1-4058-2359-3
- Mark Natola, ed. (2002). Boeing B-47 Stratojet. Schiffer Publishing Ltd. p. 46. ISBN 0764316702
- Radio Navigation & Instrument Flying. Air Pilot Publishing Ltd. 2008. ISBN 1-84336-069-1
- Jeppesen, A Boeing Company (2007). Guided Flight Discovery Private PilotJe. Jeppesen. pp. 2–66. ISBN 978-0-88487-429-4
- Compass-systems-magnetic-compass: flight-mechanic.com, Retrieved 28 January, 2020

# Permissions

All chapters in this book are published with permission under the Creative Commons Attribution Share Alike License or equivalent. Every chapter published in this book has been scrutinized by our experts. Their significance has been extensively debated. The topics covered herein carry significant information for a comprehensive understanding. They may even be implemented as practical applications or may be referred to as a beginning point for further studies.

We would like to thank the editorial team for lending their expertise to make the book truly unique. They have played a crucial role in the development of this book. Without their invaluable contributions this book wouldn't have been possible. They have made vital efforts to compile up to date information on the varied aspects of this subject to make this book a valuable addition to the collection of many professionals and students.

This book was conceptualized with the vision of imparting up-to-date and integrated information in this field. To ensure the same, a matchless editorial board was set up. Every individual on the board went through rigorous rounds of assessment to prove their worth. After which they invested a large part of their time researching and compiling the most relevant data for our readers.

The editorial board has been involved in producing this book since its inception. They have spent rigorous hours researching and exploring the diverse topics which have resulted in the successful publishing of this book. They have passed on their knowledge of decades through this book. To expedite this challenging task, the publisher supported the team at every step. A small team of assistant editors was also appointed to further simplify the editing procedure and attain best results for the readers.

Apart from the editorial board, the designing team has also invested a significant amount of their time in understanding the subject and creating the most relevant covers. They scrutinized every image to scout for the most suitable representation of the subject and create an appropriate cover for the book.

The publishing team has been an ardent support to the editorial, designing and production team. Their endless efforts to recruit the best for this project, has resulted in the accomplishment of this book. They are a veteran in the field of academics and their pool of knowledge is as vast as their experience in printing. Their expertise and guidance has proved useful at every step. Their uncompromising quality standards have made this book an exceptional effort. Their encouragement from time to time has been an inspiration for everyone.

The publisher and the editorial board hope that this book will prove to be a valuable piece of knowledge for students, practitioners and scholars across the globe.

# Index

www.ingramcontent.com/pod-product-compliance
Lightning Source LLC
LaVergne TN
LVHW080630210926
848485LV00003B/238